SYSTEMATICS AND POLLINATION OF THE "CLOSED-FLOWERED" SPECIES OF *CALATHEA* (MARANTACEAE)

SYSTEMATICS AND POLLINATION OF THE "CLOSED-FLOWERED" SPECIES OF *CALATHEA* (MARANTACEAE)

BY

HELEN KENNEDY

UNIVERSITY OF CALIFORNIA PRESS
BERKELEY • LOS ANGELES • LONDON

UNIVERSITY OF CALIFORNIA PUBLICATIONS IN BOTANY

Volume 71
Approved for publication January 7, 1977
Issued April 15, 1978

University of California Press
Berkeley and Los Angeles
California

University of California Press, Ltd.
London, England

ISBN: 0-520-09572-3
Library of Congress Catalog Card Number: 77-71064

Printed in the United States of America

CONTENTS

SYSTEMATICS AND POLLINATION OF THE "CLOSED-FLOWERED" SPECIES OF *CALATHEA* (MARANTACEAE)

by

HELEN KENNEDY

INTRODUCTION

The family Marantaceae constitutes a characteristic and conspicuous element of the understory vegetation throughout most of the lowland wet tropics. The species are pantropic, with the vast majority in the Neotropics. The species range in elevation from sea level to 1500 meters, or rarely 2000 meters, and are found in habitats ranging from gallery forest, or semideciduous forest, to cloud forest and rain forest. In general, the number of species increases with increasing moisture, decreasing elevation, and proximity to the equator. Currently, ten genera are recognized in the American tropics, the largest and most widespread being the genus *Calathea,* with approximately two-thirds of the known species. However, with more intensive exploration, especially in the Amazon basin, the number of species could easily be doubled.

The Marantaceae are a very natural group, and any member can be distinguished as such from the first seedling leaf onwards. The two vegetative characters that so delimit the group are the presence of a region of specialized cells in the uppermost part of the petiole, termed the pulvinus, and the parallel transverse veins connecting the sigmoid longitudinal veins in the leaf blade (Holttum, 1951; Tomlinson, 1961). The flower structure is equally distinctive and unique to this family. Because of this specialized flower, the Marantaceae are accepted generally by phylogenists as the most advanced family of the order Scitamineae, having a reduced androecium and reduced number of ovules. The pollination mechanism within the Marantaceae exhibits a high degree of specialization. The basic flower structure is the same for all members; thus, the differences in color, number, and size of staminodes, tube length, and opening time make them a superb group for studies of floral evolution in relation to the selection for various pollinators.

The pollination mechanism is one of explosive secondary pollen presentation. The style is held in tension by a hoodlike (cucullate) staminode, the pollen having been deposited in a stylar depression behind the stigma prior to anthesis. The style, when released by the pollinator, snaps forward, bringing the stigma in contact with the visitor and at the same time depositing the pollen upon the pollinator. Thus, each flower has but a single opportunity to be pollinated, and once the flower is tripped, there is no way for it again to be pollinated naturally. *Calathea* is well suited for pollination studies as the number of pollinated flowers in a population can be assessed readily with the aid of a hand lens; and once an artificial cross- or self-pollination is made, there is no possibility of later contamination, so that the need for bagging or emasculating the flowers is eliminated.

A previously unreported floral type in *Calathea* was observed in *Calathea marantifolia.* The flowers are tightly closed as in bud and are forced open by the pollinators.

Because the closed flower is a more specialized condition than the open flower, the potential for comparing the number, specificity, and relative efficiency of the pollinators for closed- versus open-flowered species made the closed-flowered group a logical one for concentrated study. In closed-flowered species, flowers visited by nectar thieves and those not visited at all could be censused as well as the pollinated flowers.

Since the existing taxonomic treatments of this species group were inadequate, a revision of the closed-flowered species of *Calathea* was undertaken. Since the infrageneric classification is at present in need of intensive study and revision, this grouping will be termed the "closed-flower" group of the series *Scapifoliae.* No attempt will be made to give it any formal designation as an infrageneric taxon. A more detailed study of the breeding system of these plants and their potential for crossing with related species should prove of interest, since both self-compatible and self-incompatible species are present and since one natural hybrid has been found.

ACKNOWLEDGMENTS

I am indebted to numerous individuals whose aid has furthered this study. I especially wish to express my gratitude to Dr. Robert L. Dressler for his guidance and patience throughout the project, and whose field knowledge and expertise behind the wheel of a Jeep made possible the discovery and collection of numerous species of Marantaceae, which were tolerantly allowed to fill up his orchidarium. I also wish to acknowledge his guidance in writing the dissertation and in identifying the bees involved in the study. I am indebted to the Smithsonian Tropical Research Institute for the use of their facilities as a base of operation during this project and to Dr. Robin Foster for use of his Jeep in the field and for his valuable suggestions during the study. A live collection of the species studied was maintained by Robert and Catherine Wilson at Las Cruces Tropical Botanical Garden, San Vito de Java, Costa Rica. I wish to thank Dr. Peter H. Raven for his advice and encouragement during the O.T.S. Tropical Biology course that led to the initiation of this study and for his continued interest in the project. Dr. Jesús Idrobo of the Universidad Nacional in Bogotá, Colombia, provided early instruction in the taxonomy of the Marantaceae. I thank Dr. Grady Webster for the opportunity to make observations and collections in Chiapas, México, for providing critical field supplies, and for reading the manuscript. I am grateful to Dr. Dan Nicolson for his help with nomenclatural and bibliographic problems and for reading the manuscript, and to Drs. Lorin Nevling, Donald Stone, and Robert Ornduff for their valuable suggestions and editorial comments. Drs. Herbert Baker, William Burger, Robert Read, and Beryl Simpson have read the pollination section, and Drs. Richard Holttum and Elizabeth Shaw were helpful in tracing and obtaining obscure references and publications for me. I appreciate the assistance of Cecily McNeil in proofing the manuscript.

This study has been mainly supported through an N.S.F. Graduate Traineeship and an N.S.F. Graduate Fellowship. The travel to México and Trinidad was financed by a grant-in-aid from Sigma Xi. I gratefully acknowledge the many "travel grants" provided by my parents, Elmer and Alva Kennedy.

I would like to express my thanks to the various herbaria I visited, for use of their facilities, and to those whose help and botanical knowledge considerably aided my study while there: BM, BR, CEPLAC Herbarium, COL, CR, DUKE, ECON, F, GH, K, MEXU, NY, P, RB, U, US, USJ.

I am indebted to the curators who have lent material for this study. Special thanks go to Dr. John Tucker who arranged for these loans, made space in the herbarium for my collections, and often provided guidance and advice; and to June McCaskill who unpacked and counted the various loans and maintained the records concerning them. Material from the following institutions was studied: A, B, BM, BR, C, COL, CR, DAV, DUKE, E, ECON, F, G, GH, H, HB, K, LE, M, MBM, MEDEL, MEXU, MICH, MO, NY, P, PMA, PH, PR, R, RB, S, SCZ, SP, TRIN, U, UC, US, USJ, VEN, W, WIS, and the CEPLAC Herbarium, Itabuna, Brasil, and Barro Colorado Island Herbarium, Panama Canal Zone.

METHODS AND MATERIALS

TAXONOMIC SECTION

Field Work

Field observations of the growth habit and the floral structure are necessary since these features are poorly preserved or absent altogether in dried material. The taxonomic confusion and misidentifications that have occurred in *Calathea* series *Scapifoliae* are due largely to inadequate or, in some cases, complete lack of field study. The most characteristic feature of the group, the closed flower, is not mentioned in any of the floras or monographs dealing with this group.

The majority of time was spent in Costa Rica and Panamá, where the group occurs in greatest abundance. Trips to Brasil and Trinidad were undertaken to check type localities of species whose names were being used for Panamanian plants or in the synonymy of such. Collections, measurements, or observations for the project were made at the following localities:

MÉXICO. Edo. Chiapas: Berriozabal; near Tonalá; Esperanza; and Escuintla. HONDURAS. Dept. Atlántida: Lancetilla Valley. COSTA RICA. Prov. Guanacaste: Barranca; Finca Taboga and Finca La Pacífica, both near Cañas; Tilarán. Prov. Heredia: near Puerto Viejo. Prov. Alajuela: near San Ramón and La Balsa de San Ramón. Prov. San José: Bajo La Hondura; above El General. Prov. Cartago: Tapantí; Turrialba. Prov. Puntarenas: Monteverde; Puntarenas; Agua Buena; hills above Villa Neilly; Las Cruces, near San Vito de Java; Golfito; near Rincón de la Osa. Prov. Limón: Siquirres; La Lola (Finca Waldeck); Río Hondo; Zent; Moín, Limón. PANAMÁ. Prov. Chiriquí: Volcán de Chiriquí. Prov. Bocas del Toro: Changinola; Almirante; Isla Colón; Isla Bastimentos. Prov. Coclé: El Valle de Antón; La Mesa. Prov. Panamá: Cerro Campana; Cerro Jefe; La Eneida; near Chimán; Chepo; El Llano; up the Río Maje; along Carti Highway. Canal Zone: many localities. COLOMBIA. Dept. Chocó: Quibdo; Andagoya. Dept. El Valle: near Buenaventura; Cali; Anchicayá. Dept. Nariño: Tumaco. Comm. Putumayo: Mocoa; Puerto Asis; Orito; San Juan. Intend. Caqueta: Florencia. Intend. Meta: Villavicencio; San Martín; Los Micos. BRASIL. Edo. Amazonas: Manuas. Edo. Pará: Belém. Edo. Bahia: Itabuna; CEPLAC (Cacao research station); Ilhéus. Edo. Espírito Santo: Sooretama, near Linhares. Edo. Guanabara: many localities. Edo. Rio de Janeiro: Petrópolis. Edo. São Paulo: São Paulo. Edo. Paraná: near Curitiba.

Documentation and Measurement

General.—Voucher specimens were collected at each locality where the plants were found. Both black-and-white and color photographs of each species were taken. Whenever possible, a series of photographs was taken, including one each of the habit, the inflorescence, the flower (usually showing both an opened and the normal, closed flower), and the seeds and capsule, usually with accompanying metric scale. Where mature seeds and capsules were not available at the time of collection, they were obtained by artificially pollinating plants in cultivation. The inflorescence and flowers of each species were preserved in either FPA or FAA.

For a better assessment of variation within certain species and for compatibility and crossing experiments, a number of species were maintained under cultivation at Las Cruces near San Vito de Java, Costa Rica, and at Ancon, Panama Canal Zone.

Vegetative measurements.—Measurements were made in the field on live material of all species and on recognized subspecies and varieties. The color notes were taken from live plants with one exception.

Because much critical material is not present on herbarium specimens, measurements were made *in situ*. The measurement of the pulvinus is often complicated by the acuminate extension(s) on the upper and/or lower side(s) of the petiole. The measurements given in the descriptions do not include these projections, but only the length of the area completely surrounded by specialized cells. Likewise, the petiolar measurement begins at that point. A complete set of measurements was taken of several representative individuals, and an effort was made to find the extremes and to score as many plants as possible as to number and arrangement of leaves. In small or low-density populations where the individuals were widely scattered, all plants that were encountered were measured. All individuals of *Calathea longiflora, C. vinosa, C. lasiophylla, C. venusta,* and *C. foliosa* encountered in the field were measured. In the case of *C. foliosa,* where very few individuals were encountered, and for *C. marantifolia,* where the type was an unusually small individual, herbarium material was used for additional measurements.

The size of the plant is affected noticeably by its location. Plants growing in the open, exposed to full sun, are frequently smaller at maturity than those at the forest edge or under the forest canopy.

Floral measurements.—The measurements of the inflorescence, bracts, flowers, and seeds and capsules were made predominantly from live plants. For *Calathea soconuscum, C. silvicola* ssp. *tomentosa, C. macrosepala* var. *macrosepala,* and *C. macrosepala* var. *reflexa,* additional measurements were obtained from inflorescences preserved in FAA. The differences in measurements between live flower and fruit material and that preserved in FAA were negligible in the case of the fruit. A slight shrinkage occurred in the flowers. The problems of shrinkage and distortion resulting from pressing and drying are quite pronounced in the inflorescence and particularly in flowers. Measurements made on dried flowers were 10–15% less than the actual (live) values. This difference was assessed by measuring fresh flowers, allowing them to dry, then remeasuring them. The staminodes become so fragile and distorted in drying that they are virtually useless for measurements unless rehydrated. While rehydration is occasionally helpful for study of the outer staminode, it is of questionable value for the other two staminodes, since the distortion from drying usually is not reversed satisfactorily upon rehydration.

The inflorescence diameter was measured from bract tip to bract tip across the widest portion; flowers or capsules protruding beyond the bracts were ignored. The width of the sepals, petals, and outer staminode was taken across the broadest part. Where one side of a petal is joined to the tube at a lower level than the other (as in the lateral petals), the length was measured along the midline. The height of the callose staminode was taken along the midline from the junction with the petal to the tip. The length of the cucullate staminode was measured from its junction with the callose staminode to the tip along that side.

Since the capsule is more trigonal than round in cross section, the width was measured across one face at the widest point. The aril measurements were from seeds collected at the time of, or subsequent to, capsule dehiscence, and thus, in the expanded state. Immature seeds undergo considerable shrinkage and deformation upon drying.

POLLINATION—REPRODUCTIVE BIOLOGY

Insect Observations

Photographic documentation.—Both black-and-white and color photographs were taken of the sequence of actions involved in pollination. Likewise, the robber species were photographed. The close-up photographs of the bees were taken with a telephoto lens, in combination with bellows or extension tubes, and an electronic flash.

Observations and vouchers.—Observations were made of the following species: *Calathea donnell-smithii, C. indecora, C. latifolia,* (*C. leucostachys* under cultivation), *C. longiflora, C. macrosepala, C. marantifolia, C. silvicola, C. venusta, C. vinosa,* and *C. warscewiczii* in the closed-flowered group, and *C. cylindrica, C. dressleri, C. inocephala, C. insignis, C. lasiostachya,* and *Ischnosiphon pruinosus,* which have open or partially open flowers. Upon the initiation of observations of a species at a particular site, voucher specimens of the pollinator(s) were netted and the time of day recorded. When additional time was available for observations at a particular site, the following data were taken: (1) the time of arrival of the first pollinator at the plant; (2) its behavior on approaching the inflorescence and flowers therein; (3) the total time spent by the pollinator on the flower and in the various activities involved in pollination, including opening the bud, the interval from inserting the proboscis to triggering the flower, and feeding after reinsertion of the proboscis; and (4) the number of flowers visited on an inflorescence, and the pattern (behavior) utilized in locating additional flowers and moving from one to another on the same inflorescence. Similar information was recorded for the robber species at the various localities.

Assessment of Pollination within a Population

A record was made of the time of day, the total number of flowers in an inflorescence, the number of snapped flowers with pollen, the number of snapped flowers without pollen, the number of flowers that were not visited, the number of flowers robbed, and the number that were damaged or destroyed by insects. A 14 X hand lens was used to determine the presence or absence of pollen in the concave stigmas. In the closed-flowered species, the pollen is large enough for most of the observations to be made with the naked eye.

Breeding System

Artificial pollinations were made to determine whether these species were self-compatible. Crosses between plants of the same species and a few crosses between different species were made. This was accomplished by tripping the mechanism and subsequently scraping the pollen from the style or callose staminode, then placing it in the stigmatic depression (of the same flower, another flower on the same plant, or a flower on another plant) with an insect pin. The date and location of the flower on the inflorescence were recorded. The individual flowers were tagged by knotting a colored nylon thread through one of the sepals and threading the other end through the bract. A mature capsule at the time of dehiscence is noted by the abscised calyx hanging from the bract. Capsules that are aborted before maturity can be recorded because the calyx remains affixed to these when they are shed from the inflorescence. The control flowers were tripped, but no pollen was transferred. Where only a single individual or very few individuals were involved, as in the greenhouse, the flowers were recorded but not tagged.

Nectar Measurement

The nectar at the base of the corolla tube was drawn (by capillary action) into micropipettes of known volume. The measurements were made in September at Las Cruces Botanical Garden, Costa Rica.

Germination

Seeds of *Calathea indecora, C. marantifolia, C. lasiophylla, C. latifolia, C. vinosa,* and *C. warscewiczii* were collected from dehisced capsules and planted to determine the length of time for germination.

VEGETATIVE MORPHOLOGY

Rhizome

The rhizome is sympodially branched (Holttum, 1951) and fleshy, although slightly fibrous. The internodes are often reduced; the region above each node is often appressed tomentose to sericeous.

Roots

The roots are usually thin and often quite flexible; however, those of *Calathea leucostachys* are stiff, often holding the stem and leafy portion of the plant above ground. Root-tubers are found in almost all species in this group and are usually more evident in the younger individuals. They are numerous and well developed in those species occurring in the drier, seasonally semideciduous to deciduous forest habitat: *C. soconuscum, C. macrosepala,* and *C. latifolia.* The plants die back to the rhizome each year, thus enduring the drought in a leafless condition. The storage products in the tubers, mostly starch, are utilized for rapid vegetative growth at the beginning of the following rainy season. The anatomy and development of these tubers are discussed by Weber (1957).

Leaf

Leaf blade.—One half of the leaf is rolled inward in the bud, and the other half is rolled around it. The outer half is narrower than the inner half when the leaf is mature (Holttum, 1951). The outer half frequently has ridges along its length; it represents that portion that was appressed against the midrib. Two orientations of the rolled leaves are possible: homotropic—all the leaves are rolled in the same direction; antitropic—each consecutive leaf is rolled in the direction opposite that of the previous one (see fig. 1b). The orientation of the rolled leaves, antitropic or homotropic, was used by Schumann (1902) to separate genera within the family. In the genus *Calathea,* the leaves are homotropic (Eichler, 1883).

The lateral (Holttum, 1951) or longitudinal (Tomlinson, 1961) veins are in the shape of a sigmoid curve (see fig. 1c). They are joined at right angles by a series of parallel transverse veins (see fig. 1c). Two classes of lateral veins are recognized, the main and subsidiary veins (Holttum, 1951). In the following treatment, these are referred to as major and minor veins, respectively. The majority of species in the closed-flowered group is characterized by a pattern of three minor veins between each pair of major veins. Occasionally there is but a single minor vein between the two major veins. This is commonly the case in *Calathea longiflora.* Both types of venation can be found in the same leaf. In cross section, leaves are scalloped, pleated, or ridged. In the scalloped type, the blade between adjacent veins is curved (fig. 1d). This form is found mainly in those species that occur primarily in wet or premontane forests and do not die back to the rhizome. Pleated leaves are characteristic of species from drier habitats that die back to the rhizome; however, *Calathea silvicola* also possesses this leaf form and occurs in wet or premontane forests. The pleated blade is flat between adjacent veins rather than curved (see fig. 1d). The blade of the species that die back to the rhizome is thinner in texture than the others, and the pleated conformation may confer some measure of rigidity to the blade. In the ridged type, as exemplified by *C. longiflora,* the major veins are distinctly thicker than the minor ones and appear to be raised above the level of the surrounding blade tissue, giving a markedly corrugated feeling (fig. 1d). The minor veins are usually no wider than the surrounding mesophyll.

Pulvinus.—A characteristic feature of the marantaceous leaf is the presence of a region of specialized cells in the distal portion of the petiole, termed the pulvinus (fig. 1c). "Most of the periphery of the pulvinus [is] occupied by a single (rarely two) layer of 'oblique cells' . . ." (Tomlinson, 1961, p. 69). These cells, through changes in turgor pressure, serve to orient the leaf blade. Many species hold the leaves upright in a more or less vertical position at night. The pulvinus and the sigmoidally curved lateral veins, joined by the series of parallel transverse veins of the leaf blade, are diagnostic for the entire family.

Petiole.—The petiole (fig. 1c) is often absent in those species where the leaves are clustered near the top of the stem, such as *Calathea donnell-smithii* and *C. venusta.* The petiole, when present, may bear a shallow groove along the upper side.

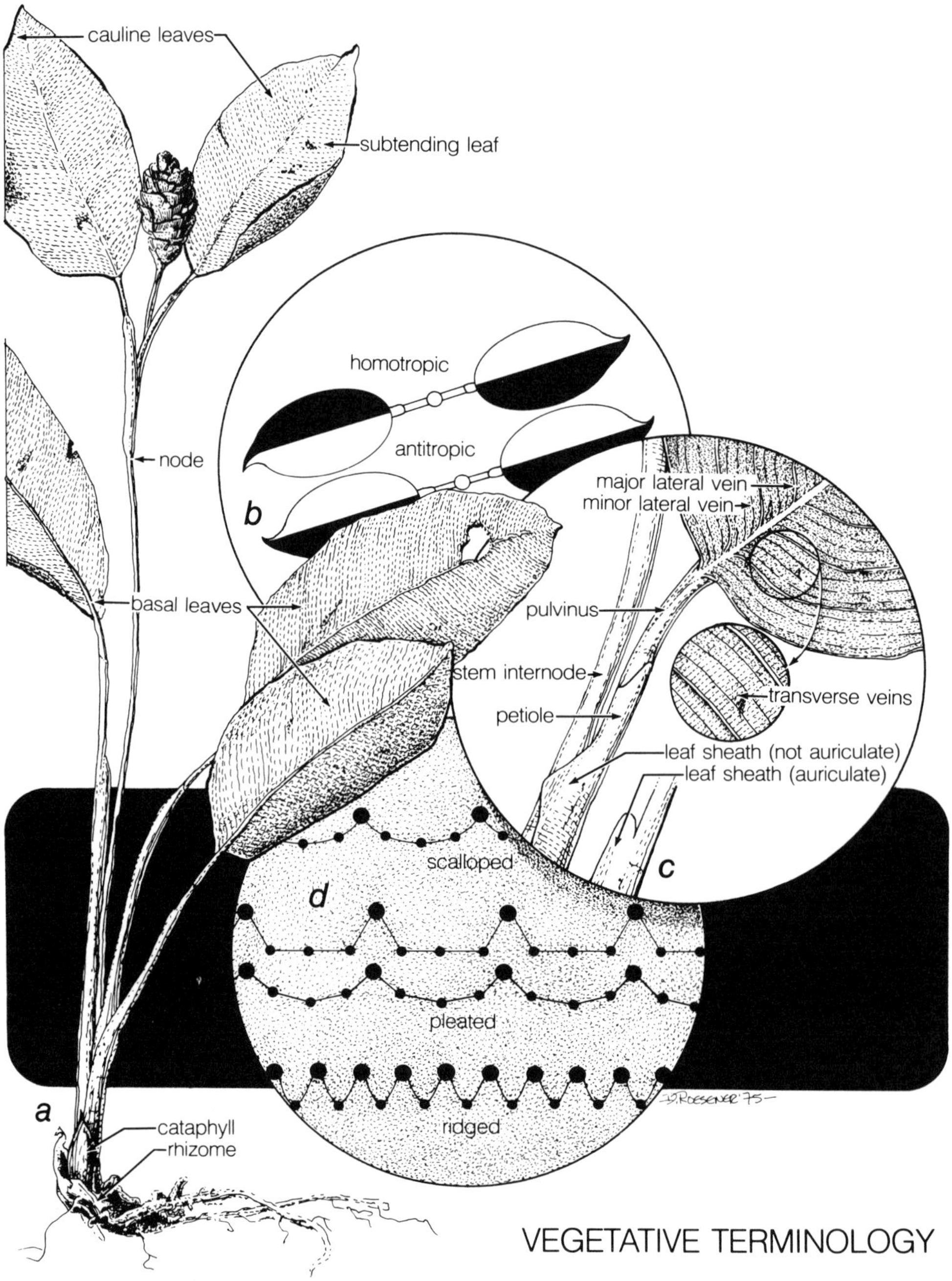

Fig. 1. Vegetative terminology. A. Habit, adapted in part from *Calathea warscewiczii.* B. Diagram of two consecutive leaves on a shoot, from above. The half of the blade in black is innermost when the leaf is rolled up, the half in white is rolled around it. Diagram after Tomlinson (1961). C. Detail of the base of the leaf, pulvinus, petiole, apical portion of the leaf sheath, and venation, showing a pair of major veins with three minor veins between them, a characteristic vegetative feature of the closed-flowered species. Note the evenly spaced parallel transverse veins between the major and/or minor veins. D. Cross-sectional view of the leaf blade showing the three types of blade morphology. The large dots represent major veins, and the smaller dots represent minor veins.

Leaf sheath.—The margins of the leaf sheath tightly clasp the stem and usually overlap throughout their length, but always overlap near the point of attachment. The sheath may be auriculate at the apex, but the apical margin does not usually extend beyond the junction with the petiole or pulvinus (fig. 1c). This can be variable within an individual plant. The lowermost leaves of *C. macrosepala* are auriculate, for example, but the upper ones are not.

Shoot

"The presence of a 2-keeled prophyll backing the primary axis, as the first leaf on every new branch, is apparently universal in the family" (Holttum, 1951, p. 255). The cataphylls, or bladeless sheaths, are the next structures to be produced on the shoot. The term cataphyll (Jonker-Verhoef and Jonker, 1957) will be used; this structure is referred to as a bladeless sheath by Holttum (1951) and Tomlinson (1961). The foliage leaves are produced later and are distal to the cataphylls on the shoot (fig. 1).

Vegetative Habit

The vegetative habit within this group is variable but follows a definite pattern consisting of an unbranched shoot bearing one to several leaves. Several shoots may arise from the sympodially branched rhizome but the aerial portion rarely branches. A shoot may form in the axil of the subtending leaf after the inflorescence dies. The young shoot first produces two to four cataphylls, after which the foliage leaves of the shoot are produced. The variation in habit results from the differential elongation of the intervening internodes. In the majority of species, the internode between the cataphylls and the first leaf (leaves) is short, and the subsequent leaves produced on the stem are separated by more or less elongated internodes. The uppermost leaf is referred to as "subtending" the inflorescence. The presence of a subtending leaf is stressed by Schumann (1902) as diagnostic of the series *Scapifoliae.* All species in which the inflorescence is terminal on the leafy shoot have an uppermost leaf that "subtends" the inflorescence. The *Scapifoliae* is not distinguished by the presence of a subtending leaf as such, but by the presence of the relatively elongate internode between the subtending leaf and the penultimate leaf.

In *Calathea macrosepala,* the internodes between the cataphylls and the first several leaves are very short, and the internode between the uppermost and penultimate leaf is elongated. The leaves, however, are more or less evenly spaced along the stem, due to the increased length of each successive leaf sheath.

A strikingly different habit based on a similar pattern of stem elongation is found in *Calathea warscewiczii.* Here, the elongated internode is usually between the penultimate and the third leaf (counting downward), and the internode between the top two leaves is quite short. The characteristic habit is of a basal "rosette" and an elongated "leafless" stem section, atop which is borne two (occasionally one or three) leaves subtending the inflorescence (see plate 20). The basal leaf sheaths are not greatly elongated and are all of similar length.

In *Calathea leucostachys,* there is a rosette-like habit. All of the internodes are more or less shortened except the one between the subtending leaf and the penultimate leaf. In some populations even this is shortened.

In several species, the internode between the cataphylls and the first leaf (leaves) is short, and the succeeding leaves are spaced out along the stem by relatively elongated internodes. This pattern is seen in *Calathea marantifolia, C. trichoneura, C. lasiophylla, C. silvicola, C. indecora, C. vinosa,* and *C. longiflora.*

Generally, in *C. indecora, C. vinosa,* and *C. silvicola,* the leaf blades are well above the level of the inflorescence, which is about half the height of the plant. In *C. longiflora* and some populations of *C. marantifolia* (plates 8, 18), the inflorescence and plant heights are nearly equal.

In *C. venusta, C. donnell-smithii,* and *C. foliosa,* the internode between the cataphylls and first foliage leaf is quite long, comprising one-half to two-thirds of the height of the plant. The internodes between the foliage leaves are short, and the leaves are clustered atop the stem. Only seedlings and juveniles of these three species have basal leaves.

REPRODUCTIVE MORPHOLOGY

Inflorescence

The pedunculate inflorescence is borne terminally on the leafy shoot. The bracts are imbricate and spirally arranged, though when pressed those of the few-bracted species may appear to be distichous. They are usually coriaceous, varying from depressed ovate or subreniform to trullate, the apex commonly rounded or broadly obtuse, rarely subacute, the margin varying from erect (straight) to revolute. Each bract subtends a number of flower pairs, usually four or more, occasionally thirteen. Within each bract, the first structure of the floral axis is an adaxial bicarinate prophyll, followed by the secondary bract opposite it, and then the flower pair itself. Each flower is subtended by a membranaceous bracteole on the side towards the prophyll (see fig. 2e). When two bracteoles are present (one for each flower), one is usually channeled and the other carinate. Occasionally, one of the bracteoles is missing, but its absence is not consistent among the flower pairs within a single bract. Its absence appears to be more prevalent in some species than others. In the axil of the secondary bract is borne another flower pair, in the axil of whose secondary bract is borne yet another, and so on, each with its concomitant prophyll, secondary bract, and bracteoles. The prophyll accompanying the second and later flower pairs is tricarinate. In the subsequent species descriptions, the term "bract" refers only to the conspicuous, coriaceous, outermost or primary bracts (Holttum, 1951). The prophylls and secondary bracts are membranaceous, transparent, or faintly colored.

Flower

The flowers (fig. 2) are decidedly asymmetrical, and the two flowers of a pair are mirror images of each other.

Sepals.—There are three, free, channeled sepals. The base and central portion of these are slightly spongy, the outer surface often shiny, and they are usually the same length as or slightly shorter than the corolla tube.

Corolla.—The corolla tube is glabrous in the upper one-fourth to one-third of its length; below this, the inside diameter of the tube gradually narrows and is puberulent within. In some species, the passage just above the nectar holding area is very

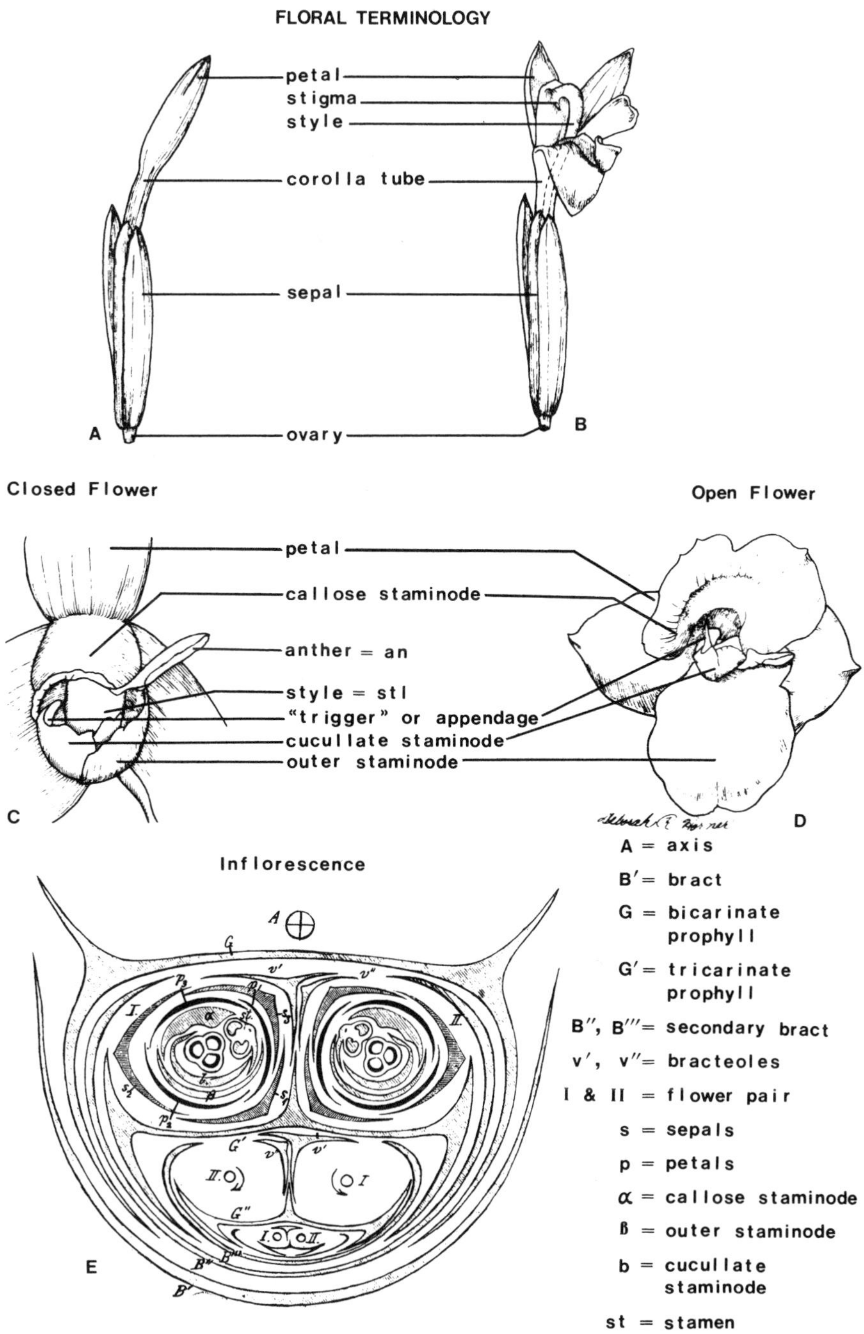

Fig. 2. Floral terminology. A. Closed flower at anthesis, prior to insect visitation. B. Closed flower after being tripped. Note curved style. C. Closed flower viewed from tip. Petals have been pulled back to show the relative position of the floral organs in an untripped flower. Note the relatively inconspicuous outer staminode compared with the open flower. D. Open-flower type at anthesis. E. Diagrammatic representation of the floral structures within one bract of an inflorescence of *Calathea*. Diagram from Eichler (1875).

constricted, due to the thickening of the wall; the inner walls are densely covered with short hairs that nearly occlude the passage through which the proboscis of the bee must pass. The lowermost portion of the tube has thin walls and is often slightly enlarged, functioning as a storage chamber for the nectar; it is *ca.* 5–9 mm. long, the length and diameter varying with the species. The corolla lobes (petals) are unequal in size. The median petal, opposing the callose staminode, is longer and narrower than the lateral two, it is elliptic, its apex is centered, and its right and left halves are symmetrical. The two lateral lobes are wider, elliptic to ovate, unequal, with the apex slightly eccentric, and the two sides are asymmetrical. At the basal junction of the two lateral petals with the median one, the petals are flush, not overlapping, and they are joined to the corolla tube at the same level. Where the two lateral petals meet, one is attached at a slightly lower level and overlaps the base of the other for *ca.* 1 mm. The apical margins of the lateral petals overlap the central one at the tip of the flower. The petals do not spontaneously reflex at anthesis as in other species of the genus, but remain together as in bud until opened forcibly by the pollinator.

Staminodes.—The lower portions of the staminodes and the stamen are adnate to the corolla tube. In *Calathea* a single stamen and three staminodes, a cucullate staminode, a callose staminode, and an outer staminode, are present. In some of the other genera, two outer staminodes are present. The outer staminode often is quite variable in size and shape within a single species or is occasionally absent altogether in some individuals. It is generally obovate or obtriangular, the apex bilobed, rounded to irregular, and the adaxial surface is slightly concave apically. The callose staminode is thick and fleshy, rectangular in shape, and concave on the inner side. It is callose throughout, bearing a small fleshy projection on one side (the side toward the stamen), against which the style rests after it has snapped forward. The margins at the apex of the cucullate (hooded) staminode are folded over and envelop the terminal portion of the style, including the stigma. Near the apex of the staminode, continuous with one of the enveloping marginal flats, is a thin, slightly curved fingerlike projection, referred to as an appendage, or sometimes as a "trigger," because of its function in pollination (fig. 2c, d).

Stamen.—The filament bears a narrow (*ca.* 1–1.5 mm. wide) lateral petaloid appendage along one side; this may be attached to the anther through a portion of its length or may end short of it. The other side of the filament, for most of its length, is fused to the callose staminode; only the terminal 1–2 mm. are free. Only one theca of the anther is fertile. The pollen grains are white, "nonaperturate, spherical, psil- and tenui-exinous; diameter 30–180μ" (Erdtman, 1952, p. 264).

Style and stigma.—The style is fused to the corolla tube up to the level of the corolla lobes and is free above that point. At anthesis it is under tension, held in place by the cucullate staminode, which also is under tension, pulling in the opposite direction (fig. 3). When the appendage, or "trigger," of the cucullate staminode is displaced, the style is released and springs upward and inward in an arc, coming to rest against the callose staminode. The style bears a shallow depression directly behind the stigma in which the pollen is borne at anthesis. The stamen and style are enveloped and pressed together by the developing cucullate staminode. The depression, in the uppermost part of the style, is perpendicular to the direction of elongation and is

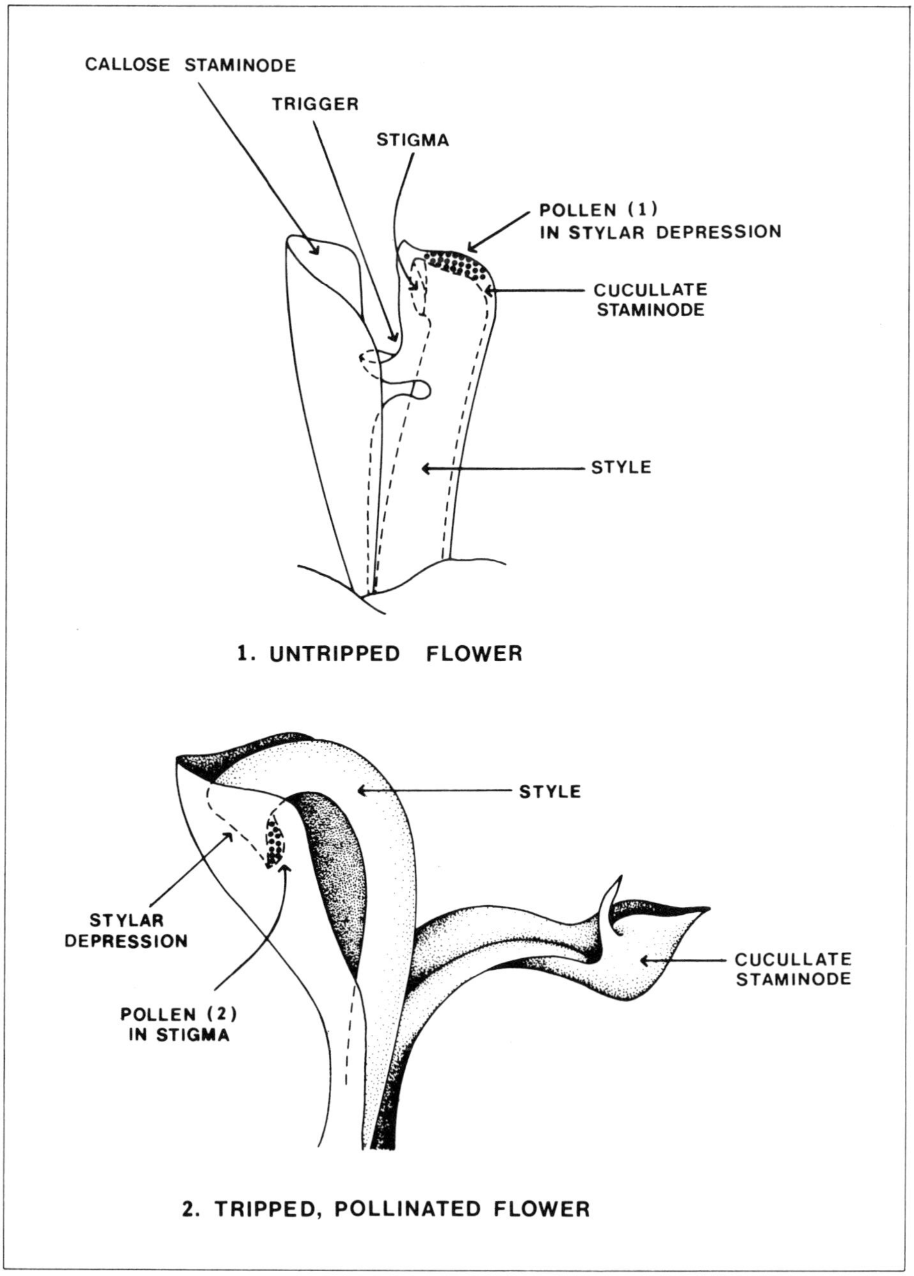

Fig. 3. Spring mechanism 1. Untripped flower: diagrammatic representation of the floral parts of a closed-flowered species showing the relative positions of the callose staminode, cucullate staminode, style, stigma, and pollen mass (1) in an untripped flower at anthesis. 2. Tripped flower: diagrammatic representation of the relationship of the floral parts after the flower has been (tripped and) pollinated. Note the curved style, reflexed cucullate staminode, and absence of pollen in the stylar depression. The pollen mass (2) that is now in the stigmatic cavity is from another flower.

positioned at the base of the anther. As the style elongates, the pressure against the dehiscing anther forces the pollen into this depression. Prior to anthesis the anther moves away from the enveloping cucullate staminode. Thus, the pollen is presented to the pollinator by the style itself. Along the leading edge of this depression is a very thin flap of tissue beneath which is produced a white viscous material, as first reported by Gris (1859). The stigmatic surface is a cup-shaped, slightly oblong, depression terminating the style. At the time of anthesis, a drop of clear liquid is found in the depression, but this usually disappears by midmorning. The style itself is hollow.

Ovary.—The inferior ovary is trilocular, with each locule bearing a single ovule. The orientation of the ovule is intermediate between anatropous and camptotrophous (Eichler, 1883). Atop the ovary are three septal glands (Gris, 1859). Pai and Tilak (1965, p. 1) state that "the nectaries are lodged within the septa and are developed right from the base of the ovary, even prior to the appearance of the ovarian loculi."

Capsule.—All Calatheas have a trilocular ovary, but occasionally all three ovules do not develop, leaving a one- or two-seeded capsule. In the closed-flowered group, the sepals remain attached to the capsule until dehiscence; they increase in size (especially at the base) and not uncommonly change color, frequently to purple. The calyx falls off at the time the capsule dehisces loculicidally. *Calathea macrosepala* and *C. latifolia* differ from the other species in that the capsule separates from the inflorescence at the time of dehiscence. The base of the capsule is swollen and mealy textured. The swelling probably aids the expulsion of the capsule, similar to the function of the aril on the seed. In the other species, the capsule is not shed until one or two days after dehiscence, and the base of the capsule is never swollen.

Seed.—The seeds are trigonal, rugose on the outer surface, slate-grey, and bear a white aril basally. Holttum (1951, p. 263) states that the arils are "turgid and assist in forcing the dehiscence of the fruit." The embryo is curved, and the perisperm canal single and straight (Gris, 1859). Humphrey (1896, p. 20) observes that "the large amount of nucellar tissue that is neither destroyed by the embryo-sac nor used for the perisperm canal becomes filled with starch to form a perisperm."

Germination falls into one of two patterns. In the more common type, germination takes place near the end of the rainy season, and the plants pass the dry season as seedlings. In the second type, the seeds are shed but do not germinate until the beginning of the following rainy season. This pattern is found in *C. soconuscum*, *C. macrosepala*, and *C. latifolia*, and appears to be an adaptation to a marked dry season.

Inflorescence Morphology—Various Interpretations

The interpretation of the pattern of the bracts, prophylls, and flower pairs within the inflorescence is controversial. Jonker-Verhoef and Jonker (1957, p. 152) state that "each [bract is] subtending a flower pair, the latter representing a dichasium of which the central flower is either suppressed or reduced to a subuliform rudiment." Holttum (1951, p. 258, 259) observes that "in simple spikes . . . each bract contains a condensed monochasial cyme bearing a few flowers. . . . Each successive branch of the cyme ends in a pair of flowers." However, Holttum later mentions that "Schumann reports that the rudiment of a third flower between the two is sometimes found;

we may thus perhaps interpret this rudiment as the real apex of the group, and the pair of flowers as lateral, as in the ultimate branchings of a dichasial cyme" (Holttum, 1951, p. 260). This interpretation assumes that the subuliform rudiment is always single in its occurrence, which it is not. A discussion of the cymose theory with especial regard to the work of Eichler, together with Thompson's own theory is given by Thompson (1933, p. 34). He observes that "with each conceivable modification of the view of cymosity difficulties of definite magnitude arise. . . ." A detailed study of the ontogeny of the floral branch system is necessary to resolve this controversy.

Floral Morphology—Various Interpretations

The most prevalent interpretation of the staminal whorls is that the fertile stamen, callose staminode, and cucullate staminode represent the inner whorl, while the petaloid outer staminode is part of the outer whorl (Nees, 1831; Gris, 1859; Eichler, 1883; Holttum, 1951; Tilak and Pai, 1966, 1968). In his comparison of the marantaceous flower with that of *Canna,* Costerus (1918) suggested that the lateral petaloid appendage of the stamen represents the third staminode of the inner whorl, along with the callose and cucullate staminodes, and that the stamen is actually part of the outer whorl. Tilak and Pai (1966), studying the vascular traces, substantiate the view that the stamen is of the inner whorl.

TABLE 1
Terminology Used Herein and Comparable Terms of Other Authors

1. Bract.	Primary bract (Holttum, 1951); spathe (Thompson, 1933); Deckblatt, Bracteen (Schumann, 1902); Deckblatt (Eichler, 1883); brattea comune, gluma (Delpino, 1869).
2. Bicarinate prophyll.	2-keeled prophyll (Holttum, 1951); twin-keeled bract (Thompson, 1933); zwei-kielig Vorblatt (Eichler, 1883; Schumann, 1902); 2-kielig Grundblatt (Eichler, 1875 and 1883); brattea bicarinata, glumella (Delpina, 1869); bracteoles (Hooker, 1865).
3. Tricarinate prophyll.	3-keeled prophyll (Holttum, 1951); 3-kielig Vorblatt or Grundblatt (Eichler, 1875); partial bracteae (Lindley, 1819).
4. Secondary bract.	Simple bract or mesophyll (Holttum, 1951); bractlike bodies (Thompson, 1933); Zwischenblatt or mesophyllis (Schumann, 1902); Deckblatt (Eichler, 1875); Zwischenblatt (Eichler, 1883).
5. Bracteole.	Bract or bracteole (Thompson, 1933); bracteolen (Schumann, 1902); Vorblatt (Eichler, 1875).
6. Claviculate bracteole.	Bracteoles (Holttum, 1951); subuliform rudiment (Jonker-Verhoef and Jonker, 1957); bracteolis indurato-claviculatis (Schumann, 1902); Vorblatt (Eichler, 1883).
7. Outer staminode.	Outer staminode (Holttum, 1951); staminodium exterior (Schumann, 1902); Aussenstaminod (Schumann, 1902); Flugel (Costerus, 1918); parapetalum (Horaninow, 1862); labellum (Lindley, 1826b).
8. Callose staminode.	Fleshy staminode (Holttum, 1951); staminodium callosum, Schwielenblatt (Schumann, 1902); labellum (Tilak and Pai, 1966); Labell (Nees, 1831); staminode calleux (Gris, 1859).
9. Cucullate staminode.	Hooded staminode (Holttum, 1951); staminodium cucullatum, Kappenblatt (Schumann, 1902); Kapuzenblatt (Eichler, 1883); staminode cucullé (Gris, 1859).
10. Petaloid appendage.	Petaloid appendage (Holttum, 1951); Kuppler (Costerus, 1918).

POLLINATION—REPRODUCTIVE BIOLOGY

Mais pourquoi y a-t-il un staminode cucullé et un staminode calleux? à quoi servent-ils? Pourquoi ce brusque mouvement du style? Pourquoi le stigmate vient-il appliquer sa tête à la face interne du staminode calleux? Questions bien dignes d'exciter la curiosité du botaniste, auxquelles personne, que je sache, n'a jusqu'ici répondu. . . .

Gris, 1859, p. 210

EARLY OBSERVATIONS ON FLORAL BIOLOGY OF MARANTACEAE

Botanists have always had a certain fascination with the unusual floral apparatus of the Marantaceae and with its significance in reproduction. Many species of Marantaceae were collected and cultivated for their strikingly beautiful leaves. In the course of describing new species in cultivation, the botanists often commented on the peculiar floral structures and hypothesized as to their nature and function. Lindley (1819), who was one of the first to give a detailed account of the floral parts and their ontogeny, described the mode of pollen transfer within the bud from the stamen to the stylar depression. Later, in describing another species, he (Lindley, 1826a) reported on the spring action of the style.

Nees (1831) compared the floral structure of *Maranta arundinacea* and *Thalia dealbata* with that of *Canna.* He was, however, apparently unaware of the significance of the curved style and erroneously attributed its shape to asymmetrical development, noting that there were numerous examples of these "Missgestaltungen."

In 1859 Gris published the results of his excellent and detailed observations on the flowers of various Marantaceae growing in the greenhouse of the Museum Nacional d'Histoire Naturelle, Paris. In addition to the transfer of the pollen into the stylar depression and the spring action of the style, Gris also reported the presence of a glandular region along the leading margin of the stylar depression, which separates it from the stigmatic cavity. He referred to this glandular region as an "appareil accessoire, *collecteur.*" Gris realized the need for external agents to trigger the mechanism and thus effect pollination. He stated (p. 211), "Les vents, les oiseaux, les insectes, sont sans doute, pour ces plantes bizarres, les intermédiaires naturels indispensables à l'impregnation."

Hildebrand (1870a) made essentially the same observations as Gris regarding pollen transfer and the need for an external agent to displace the cucullate staminode and thus release the style. Contrary to Gris, Hildebrand believed that the function of the spring style and the pollen placement was to prevent autogamy and postulated that the pollen was transported by the same organism that caused the release of the style. He also postulated that, upon release of the style, it sprang forward with the stigma first scraping against the insect's body, thus collecting the pollen brought from a previously visited flower, and then depositing in the same place on the insect its own pollen (Hildebrand, 1870b).

Delpino (1869) made somewhat more detailed observations. He noted the unequal development of the tension in the style and in the *callo* portion of the cucullate staminode, which are the basis of the explosive mechanism. He stated that an insect putting its head into the flower in search of nectar, touches the trigger, or *il dente,* with its proboscis, releasing the style, which strikes the insect's sternum or abdomen. By this means the glandular substance (bordering the stylar depression) is smeared on the insect's body, and the pollen deposited therein. Unlike Hildebrand, Delpino had been able to observe actual insect visits. Müller (1883) and Knuth (1899) both present summaries of the work of Hildebrand and Delpino.

Ducke's (1901) observations on the bee fauna of Pará, Brasil, constitute one of the first published accounts of visits to Marantaceae by Euglossine bees. Dodson and Frymire (1961), Vogel (1963, 1966), Dodson (1966), and Dressler (1967, 1968a, 1968b) mention the visits of Euglossines to Marantaceae. A good summary of the published accounts of various species of Euglossine bees that feed on Marantaceae was given by Zucchi *et al.* (1969).

EUGLOSSINE BEES

Euglossine bees are the major group of pollinators for the New World Marantaceae. The genera *Euglossa* and *Eulaema* are most commonly encountered as pollinators, especially in the genus *Calathea.* The Euglossine bees are in the family Apidae along with *Apis mellifera* (the honeybee) and *Bombus* (bumblebees), but in a separate tribe, *Euglossini.*

The tribe *Euglossini* constitutes a relatively advanced group, though they are solitary or communal and lack the complex social structure found in other members of the family (Dressler, 1968a; Bennett, 1965). They are found only on the American continent and are mainly restricted to the tropics (Moure, 1967; Dressler, 1968a). Six genera are currently recognized within this tribe: *Euglossa, Euplusia, Eufriesea, Eulaema, Aglae,* and *Exaerete.* All genera but *Eufriesea* and *Aglae* have been reported to visit Marantaceae. *Euglossa* and *Eulaema* figure most importantly in Marantaceae pollination. Of the Euglossas, only the subgenus *Glossura,* comprising "the largest species with very long labio-maxillar complex . . ." (Moure, 1967, p. 397), figure in the pollination of the closed-flowered group. The pollinators of the closed-flowered Calatheas are in both subgenera of *Eulaema,* but predominate in the subgenus *Apeulaema* (three species), with only one, *E. speciosa,* in the subgenus *Eulaema.*

BEHAVIORAL ASPECTS OF POLLINATION

Sequence of Pollination of Closed-Flowered Species

Nearing the inflorescence, the bee may circle it or may approach it on one side at the level of the basal bracts, proceeding toward the apex, then flying part way around before beginning again near the base and proceeding upward in search of mature flowers. When a mature flower is encountered, the bee approaches it more closely, flying in a tight figure-eight pattern in front of the flower, with the antennae held forward and slightly downward. Just prior to alighting, the bee hovers directly in front of the flower and touches it briefly with the antennae (plate 2a). Then, gripping the sides of the flower near the tip with the forelegs and then midlegs, the bee swings downward from the more or less horizontal hovering position, then grasps the flower with the hind legs. Holding on to the flower with the midlegs and hind legs (the hind legs appear to be bracing against the flower rather than grasping it), it then uses its mandibles to chew the tip of the flower where the three petals meet, thus facilitating the later separation of the petals (plate 2b). Following this, it proceeds to pull the lateral petals apart with the front tarsi. The bee then pushes its head under the median petal and forces it upward and back, away from the callose staminode (plate 2c). It then rears away from the flower, pivoting on the hind legs with the mid pair somewhat extended, brings the proboscis forward, and inserts it between the callose and the cucullate staminodes. The curved inner surface of the callose staminode thus channels the proboscis into the corolla tube. Just before the bee's head enters the flower, the proboscis is fully extended, exposing the proboscidial fossa. The bee then usually begins feeding, but soon tries vigorously to force its proboscis farther into the flower. The forelegs are usually folded against the body at this time, the bee using only the midlegs to pull itself forward into the flower. Its head is now positioned between the cucullate staminode (below) and the callose staminode (above). As the head is forced into the flower, the side of the head (malar space) touches the "trigger" (appendage) of the cucullate staminode, depressing it laterally, thus releasing the style. The style arches upward violently, throwing the bee forward and momentarily pinioning its head against the callose staminode. As the style curves forward, it is precisely positioned so that the terminal cup-shaped stigma enters the proboscidial fossa, scraping the pollen previously deposited there into the stigmatic cavity. In the same motion, the style places its own pollen in the same area. The bee then pulls its head from the style (which usually continues to bend until it is curled in upon itself, resting against the thick, flaplike protuberance of the callose staminode), reinserts its proboscis, and resumes feeding (plate 2d, e).

Sequence for Open Flowers

In open flowers, the mode of pollination is basically the same as for the closed-flowered species, except that the first step of opening the flower is usually omitted. As with the closed-flowered species, *Euglossa imperialis* was observed to hover, flying in a tight figure-eight pattern, in front of the (open) flower before alighting on it. The Euglossas, being relatively small and light, frequently grasp the outer staminode when landing, subsequently bringing the proboscis forward and inserting it directly into the corolla tube. Occasionally, Eulaemas (and some of the largest Euglossas) will grasp the bracts subtending the flower (especially in the distichously bracted species such as *Calathea insignis*). After inserting its proboscis, the bee, attempting to get more nectar, forces its head into the flower, tripping the mechanism. In *Ctenanthe dasycarpa*, the proboscis itself triggers the flower, and the pollen is placed on the upper surface of the proboscis.

In Panamá, *Euglossa imperialis* has been observed to open and pollinate unopened flowers of *Ischnosiphon pruinosus* as well as to pollinate the usual spontaneously opened flowers. *Eulaema speciosa* frequently opens the unopened flowers of *Calathea inocephala, C. guzmanioides,* and *C. lasiostachya.* These species are opened readily in a few seconds, since in most cases the petals rapidly reflex upon being disturbed. *Ischnosiphon pruinosus* occasionally requires as long as 13 seconds to be opened.

Additional Observations and Slight Variations in the Behavioral Sequence

Approach to Inflorescence.—Where few plants of *Calathea marantifolia* were present, *Eulaema speciosa* often approached the inflorescence rather directly. In denser populations, both *E. speciosa* and *E. cingulata* would often fly back and forth through the clumps several times before investigating a particular inflorescence. When approaching a lone individual of *Calathea inocephala, Eulaema speciosa* flew toward the inflorescence, then away 1.5–2 m., repeating this several times before settling on the inflorescence. Whether this was a response to my presence or an unusual behavioral pattern is not clear, though it was observed on several occasions.

Approach to flower.—*Eulaema speciosa* was able to discriminate between fresh flowers and day-old flowers of *Calathea marantifolia*; unvisited flowers from the previous day remained untouched while the fresh flowers on the same inflorescence were pollinated. Often the bees (*Eulaema speciosa, E. cingulata,* and *Euglossa imperialis*) leave the proboscis extended after quitting one flower and flying to the next on the same inflorescence, but retract it just prior to alighting on the closed flower. In the case of the open flower (*i.e., Calathea insignis, C. lutea, C. lasiostachya, Ctenanthe dasycarpa,* and others), the proboscis is usually not retracted since the extended position allows for immediate insertion in the corolla tube upon alighting. In the case of *Ctenanthe dasycarpa,* this is especially noticeable since only the proboscis enters the flower.

Opening of the flower.—Occasionally, through abnormal development, the corolla is less elongated than usual beyond the calyx. This renders opening the flower much more difficult, at times impossible. The bee will generally spend two to three times the normal opening time working on such a flower before abandoning it. *Eulaema speciosa* spent 50 seconds trying to open a malformed bud of *Calathea marantifolia,* after which an additional 49 seconds were expended attempting to insert the proboscis into the flower. The corolla in this case was only exerted 11 mm. beyond the sepals, as opposed to 23 mm. in the normal condition. Another individual of *E. speciosa* worked at a bud for 30 seconds without success before flying off. After some difficulty in opening a flower of *C. marantifolia, E. speciosa* began to claw at the petals until an opening had been torn in the side of the petal, through which the proboscis was inserted into the flower. This form of behavior is probably rather rare and was observed only once.

Where a flower has been opened slightly but not snapped by a robber species (plate 7e), a legitimate pollinator may subsequently visit it and effect pollination. This was evident from a census (p. 22) of the flowers taken at various times throughout the morning. The census showed a decrease in the number of visited, but unpollinated, flowers through the course of the morning, and a concomitant increase in the number of pollinated flowers. If the inflorescence is not oriented in the normal upright position, because the plant is bent or falling, the bee, subsequent to landing, will assume the normal positional relationship to the flower (*i.e.,* grasping the lateral petals) and proceed to open and pollinate it.

Inserting the proboscis.—Inserting the proboscis is rarely difficult for the bee, usually taking only a few seconds. However, a rather battered male *Eulaema speciosa,* visiting *Calathea lasiostachya,* appeared to have considerable trouble finding the entrance to the corolla tube. He made various attempts at the flower for several seconds before correctly inserting the proboscis. A female *E. speciosa,* trying to insert the proboscis into a flower of *C. lutea* with some force, caused the maxillae to separate at the distal end of the proboscis.

Feeding.—After the flower has been snapped, the cucullate and callose staminodes are well separated, enabling the bee to get its head very close to the entrance of the corolla tube (plate 2f). Thus, after snapping, the "effective" tube length (for the bee) is shortened. The usual orientation for feeding is with the bee grasping a lateral petal with the hind legs, with one midleg resting atop the callose staminode and the other midleg on the same lateral petal as the hind legs or on the remaining one. The forelegs are often folded against the body at this time. Depending on which of the two lateral petals the bee uses to support itself, two orientations are possible (plates 2f, 3a).

NECTAR ROBBING

Any visit to a flower in which the visitor takes nectar (or pollen) but does not pollinate the flower is considered robbing. Commonly, long-tubed flowers are punctured, or a hole is chewed into them near the base of the corolla tube, thus allowing access to the nectar.

In the closed-flowered species of *Calathea,* the sepals are relatively tough and coriaceous, and are usually as long as, or only slightly shorter than, the corolla tube. The lower portion of the tube is well within the subtending bract and is also protected by the surrounding bicarinate prophylls and secondary bracts. While this affords protection for the nectar from most insects, hummingbirds (*Phaethornis superciliosis* L. ssp.) were observed to puncture through the calyx of *Calathea marantifolia* rather easily. Bees of the genus *Trigona* occasionally cut a small hole in the corolla tube of *C. latifolia* just above the calyx; however, the bee is too large to enter the corolla tube and could not possibly reach the nectar through the hole at that level. The significance of such activity is unclear.

Insects that have a long, thin proboscis can often take nectar from shorter-tubed flowers without either their bodies or the proboscis coming in contact with the pollen and/or stigmas. This form of robbing occurs in open-flowered Calatheas such as *C. insignis,* where moths, butterflies, and long-tongued bees have often been seen probing for nectar.

In the closed-flowered group of *Calathea,* specialized forms of robbing take place involving the bees of the larger-sized species of *Eulaema* (in subgenus *Eulaema*) and less frequently the longest-tongued Euglossas (subgenus *Glossura*), as described below.

Behavioral Aspects of Robbing

Eulaema meriana *type.*—The relatively long tongues of *Eulaema bombiformis* and *Euglossa asarophora* enable them to exploit several of the closed-flowered group in a manner similar to the robbing by lepidopterans. The approach to the inflorescence and flowers is basically the same as for pollination. The bee alights in a like manner, grasping the lateral petals, and chews at the tip of the flower where the three petals meet. The bee pulls the lateral pedals apart at the tip and also forces the median petal away from the callose staminode with its head. The bee now extends the proboscis, inserts it into the flower, and begins feeding; the entire sequence to this point paralleling that of pollination. While the pollinator would then force its head farther into the flower, thus tripping the mechanism, the robber species simply feeds, the head coming no closer than the apex of the callose staminode (plate 1). The effect of corolla tube length in determining whether a species will serve as pollinator or robber is well illustrated by the fact that *Eulaema meriana* robs *Calathea latifolia* and *C. marantifolia* (both having a mean tube length less than 30 mm.), while it pollinates *C. donnell-smithii* (mean tube length *ca.* 40 mm.). That this is not the sole determining factor is emphasized by the fact that *Euglossa asarophora* is the pollinator of *C. longiflora* but has been observed to rob *C. donnell-smithii* (which has an equally long, if not longer, corolla tube).

Eulaema seabrai *type.—Eulaema seabrai* has a somewhat shorter tongue and robs in a quite distinctive manner. The bee alights on a closed flower (of *Calathea*) and proceeds to open the bud by the same methods as *E. meriana* or *E. speciosa.* After the proboscis is inserted into the flower, the bee does not force the head directly into the flower but eases it to one side, keeping the head entirely outside of the petals and allowing the proboscis to pass between the adjoining margins of the median and lateral petals. The bee is oriented on its side in relation to the flower, a position similar to that of *E. speciosa* in plate 2f, except that the flower remains unsprung. The bee forces the proboscis downward between the petals to the top of the corolla tube. The bee occasionally accomplishes this without triggering the style. However, the trigger (fig. 2b, c) is on the side toward the bee and can be tripped by the bee's proboscis. In any event, the bee's head is well out of the way should this occur; hence, the pollen of the flower is wasted.

The behavioral pattern was observed for *Eulaema seabrai* when robbing both *Calathea marantifolia* and *C. macrosepala* at Monteverde (northwestern Costa Rica), *C. marantifolia* near San Vito de Java (southwestern Costa Rica), and *C. latifolia* and *C. marantifolia* in central Panamá.

At San Vito de Java, an individual of *Eulaema seabrai,* after inserting the proboscis and orienting itself sideways to the flower, proceeded to "claw" at the semiopened apical portion of the flower with its midleg causing the flower to be sprung. The bee immediately assumed the "normal" (plate 2f) feeding position. This behavior was repeated at subsequent flower visits by the same bee. It appeared to be a learned variation on the general pattern since it was not observed at all localities.

Discussion

Brian (1957) mentions differences between *Bombus* species in their tendency to rob and discusses the controversy over whether these differences are innate or learned. She refutes some of the evidence previously used in support of both sides of the question but does indicate some support for an innate behavioral difference. "Thus while [*Bombus*] *hortorum* appears conservative and stereotyped, always visiting flowers by the obvious and correct entrance, [*B.*] *lucorum* is an opportunist obtaining nectar by a variety of methods" (Brian, 1957, p. 79). Certain strains of honeybee

(*Apis mellifera*) have been observed to persistently pollinate alfalfa flowers in the correct manner despite being hit repeatedly by the style and anthers that spring forward upon release (Robin Thorp, pers. comm.). Such evidence seems to support a genetic (*i.e.,* innate) component to such behavior. Though the definitive observations are still to be made, there is reason to believe that this may be the case with *Eulaema speciosa* and *E. seabrai.* Only once in 138 observations of *E. speciosa* visiting closed flowers was it observed to rob rather than pollinate the flower. In all observations of *E. seabrai* on closed flowers (in México, Costa Rica, and Panamá), the basic manner of robbing was consistently the same (with the exception of a slight modification noted previously). This species was never once observed to pollinate a flower. *Eulaema meriana* likewise exhibited a consistent pattern in its robbing behavior, both in northern and southern Costa Rica, and in central Panamá. In order to decide the question (of innate *vs.* learned behavior) with regard to *E. seabrai,* it would be necessary to observe newly emerged, naive individuals of *E. seabrai* on their first visit to a closed flower. *Eulaema meriana* has been observed to legitimately pollinate *Calathea donnell-smithii* (a relatively long-tubed species), although more observations on the behavior of both *E. meriana* and *E. bombiformis* in relation to the longer-tubed closed flowers would be quite informative.

The frequency of robbing visits to *Calathea marantifolia* was low compared to pollinator visits when the plants were in low-density populations of scattered individuals (fig. 4). In a large population of *C. marantifolia,* such as in a garden, the frequency of robbing rose considerably since the pollinator population was unable to utilize such a large number of flowers. The same phenomenon was observed with a large population of *C. latifolia* on Ancon Hill and Pipeline Road, Canal Zone, and with a large population of *C. marantifolia* growing in a pasture area near Monteverde, Costa Rica. In low-density populations of these closed-flowered species, the potential detrimental effect of robbing is lessened by the fact that the legitimate pollinator visits the plant earlier than the robber species (fig. 4).

Robbing by *Eulaema meriana* and *E. bombiformis* is only marginally detrimental since the flower is not tripped and can be (and frequently is) subsequently pollinated. A census at 5:30 A.M. of *C. marantifolia* (20 June 1969, near San Vito de Java) showed three inflorescences to have a total of six flowers that had been robbed. A second census at 10 A.M. showed all these robbed flowers to have been pollinated. On the other hand, *E. seabrai* is potentially detrimental to the reproduction of the plant species, and should its frequency of visitation to a particular species greatly increase, selection for a "defense" against it might be expected.

SPECIFICITY

The "primary attractant" (Faegri and van der Pijl, 1966) of the closed-flowered species and, indeed, probably of all Marantaceae, is the presence of nectar. In *Calathea, Ischnosiphon,* and *Monotagama,* the pollen is placed in the proboscidal fossa, an area that the bee cannot reach in cleaning itself. Van der Pijl (1960) noted that in the orchid family, where a similar all-or-none mechanism prevails, the pollen is removed completely from the "food-sphere."

According to the system of Faegri and van der Pijl (1966), the closed flowers would be classified as euphilic/monophilic, being well adapted to their pollinator and being

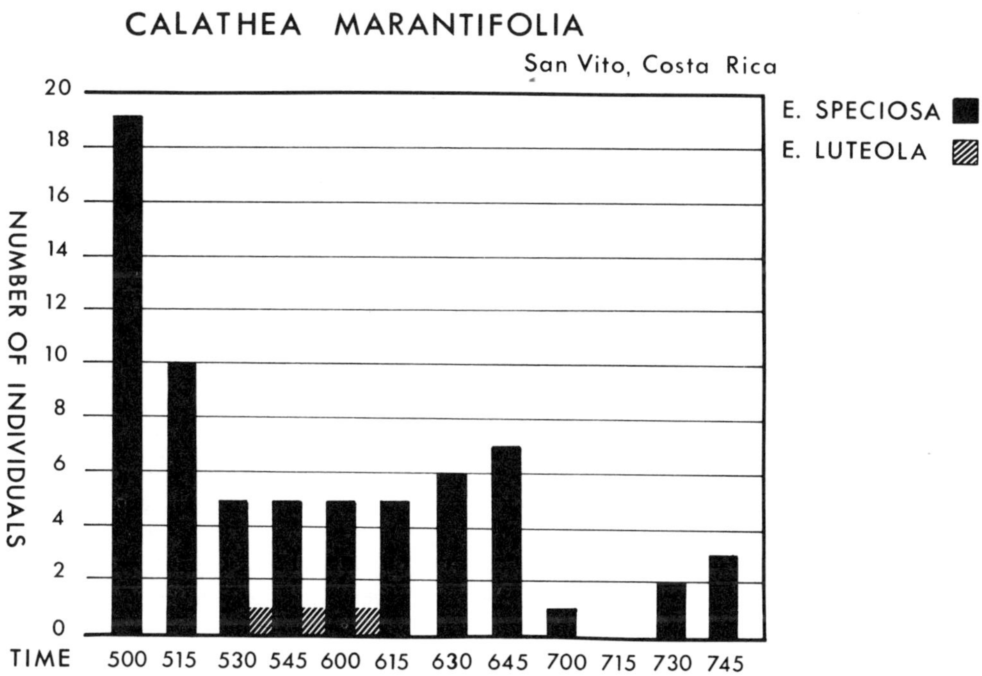

Fig. 4. The number of pollinators (*Eulaema speciosa*) or robbers (*Eulaema seabrai*) observed visiting *Calathea marantifolia* from 5:00 A.M. to 7:45 A.M. at San Vito de Java, Costa Rica. The number of individuals is the number observed over a period of a month and a half. Each time given represents the band of time from that figure to the next figure.

pollinated by a single species or a few closely related species. Though the bees are well adapted to them, these Calatheas probabiy are not their main food source.

The specificity is rather one-sided, *i.e.*, the plant is monophilic while its pollinator is polytrophic. This relationship is similar to that for the Euglossine-pollinated orchids though usually not so specific. Baker and Hurd (1968, p. 403) noted that in general the pollen-gathering females exhibited a greater specificity in the choice of flowers from which they obtain pollen, whereas "the least specificity involves the visitation of flowers by either male or female bees for nectar. . . ." Dressler (1968b, p. 202) remarked that "many of the plant species which have been observed are visited and pollinated by a single [Euglossine] bee species, though the bee is likely to visit and pollinate other unrelated or distantly related plant species." The variety of host plants can be rather great for a particular species of bee (Dodson, 1966). Of the nine species observed (table 2) in their native habitat, four were visited by only one species of pollinator, two by two species of pollinator, two by three pollinator species, and one by five pollinator species (table 4). In general, at any specific locality at a particular time, only one pollinator species was observed. The occurrence of two (or more) pollinators, simultaneously visiting the same species, is almost always correlated with the occurrence (frequently unnatural) of high plant-population density, with the plants often in dense stands. Janzen (1971) commented that a female Euglossine probably has five to ten species on her feeding route. I have often observed

TABLE 2
Pollinators and Robbers of Closed-Flowered Species

Species	Pollinator	Robber	Locality
C. macrosepala var. *macrosepala*	*Eulaema cingulata* *E. polychroma*	*Eul. seabrai*	MÉX. Chiapas, near Tonalá
	E. cingulata *E. polychroma*	*E. seabrai*	CR. Puntarenas, Monteverde
*C. macrosepala** var. *reflexa*	*E. cingulata* *E. speciosa***	*E. seabrai*	CR. Puntarenas, San Vito, Las Cruces
C. latifolia	*E. nigrita* *E. cingulata*	*E. meriana*	PAN. CZ. Ancon
	E. cingulata *Euglossa imperialis*	*E. meriana* *Eul. seabrai*	PAN. CZ. Pipeline Road
	Eug. imperialis	*Eul. meriana*	PAN. CZ. BCI
	Eul. cingulata	*Euglossa* sp.	TRIN. near St. Augustine
C. marantifolia	*Eul. polychroma* *E. cingulata*	*Eul. seabrai* *E. meriana*	CR. Puntarenas, Monteverde
	E. cingulata		CR. Cartago, Turrialba
	E. speciosa	*E. meriana* *E. seabrai*	CR. Heredia, near Pto. Viejo
	E. speciosa *E. cingulata***	*E. seabrai* *E. meriana***	CR. Puntarenas, San Vito, Las Cruces
	E. speciosa	*E. seabrai*	PAN. Pan. Cerro Campana
	E. cingulata		PAN. CZ. Pipeline
	Eug. imperialis		PAN. CZ. BCI
	Eul. speciosa *Eug. ignita*		PAN. Colón, Río Buenaventura
C. vinosa	*Eug. asarophora*		CR. Puntarenas, Osa peninsula & Golfito
*C. vinosa**	*Eul. speciosa*		CR. Puntarenas, San Vito, Las Cruces
C. indecora	*Eug. ignita* *Eug. imperialis*		PAN. Bocas del Toro, Almirante
C. silvicola ssp. *silvicola*	*Eulaema speciosa*		PAN. Pan. Cerro Campana
	E. speciosa		PAN. Pan. Cerro Jefe
C. venusta	*E. speciosa*		PAN. Colón, Río Guanche
C. donnell-smithii	*E. speciosa* *E. meriana***	*Euglossa asarophora*	PAN. Pan. Cerro Jefe
	E. speciosa		PAN. Pan. Cerro Campana
C. longiflora	*Eug. asarophora*		CR. Puntarenas, Osa peninsula
*C. leucostachys**	*Eul. speciosa*		CR. Puntarenas, San Vito, Las Cruces

*Observations under cultivation, not in native habitat.
**Occasional, of minimal importance at that site (especially as compared with the other species mentioned).

TABLE 3
Pollinators and Robbers of Closed-Flowered Calatheas

Pollinators	Robbers
Eulaema Speciosa	*Eulaema meriana***
E. cingulata	*E. seabrai*
E. nigrita	*E. bombiformis*
E. polychroma	
*Euglossa asarophora**	
E. imperialis	
E. ignita	

* One observation of robbing *C. donnell-smithii.*
**One observation of pollinating *C. donnell-smithii.*

TABLE 4
Specificity of Pollinators of Closed Flowers

No. of Pollinators Observed per Plant Species*	No. of Plant Species with Observed No. of Pollinator Species	Plant Habitat
1	4	Wet, deep forest
2	2	Wet forest, 1 forest, 1 semiopen to forest
3	2	Semideciduous forest
5	1	Moist to wet forest, forest-to-forest edge, occasionally in semiopen areas, widespread, variable

*Usually only one main pollinator per plant species at any single location.

females on marantaceous flowers bearing *Costus* pollen on their back and pollen of a different plant in the corbicula.

Influence of Ecological Factors on Pollinator Species

An important factor determining the pollinator of a plant species in a particular area is the degree of overlap between the habitat preferences (or tolerances) of the pollinators and the plants.

Moisture regime.—Calathea latifolia and *C. macrosepala,* which are frequently in open areas and grow in the driest habitat in which the closed-flowered species occur (the semideciduous forest), were observed to be pollinated by *Eulaema cingulata* in six of the seven localities studied. At the driest end of its range, *C. marantifolia* occurs in semideciduous forest along Pipeline Road and on Barro Colorado Island together with *C. latifolia.* (For more complete locality data, see table 2). Both *C. marantifolia* and *C. latifolia* are pollinated by *E. cingulata* at the Pipeline Road locality, whereas *E. speciosa* was not observed. However, on Ancon Hill, a drier locality, there is a shift to *E. nigrita* as the main pollinator of *C. latifolia. Eulaema cingulata,* which has a broad ecological range (Robert Dressler, pers. comm.), also was found to pollinate

C. marantifolia in the premontane rain forest of San Vito de Java and the premontane wet forest at Turrialba. *Calathea marantifolia* is pollinated mainly by *E. speciosa,* a bee with a narrower ecological range, most commonly found in the wetter forest habitats. Of the eight localities where *C. marantifolia* was observed (table 2), *E. speciosa* figured as the main pollinator in four: Puerto Viejo, Buenaventura, San Vito de Java, and Cerro Campana. The first two are lowland wet forests and the second two, premontane wet or cloud forests. *Eulaema speciosa* was observed to pollinate *C. macrosepala* var. *reflexa,* which was growing in a garden near San Vito de Java, a wetter environment than that in which this variety occurs naturally. *Calathea silvicola, C. venusta,* and *C. donnell-smithii* are restricted to lowland wet, or premontane wet, to cloud forest habitats, and all are pollinated by *E. speciosa. Euglossa asarophora,* a bee of lowland wet to cloud forest habitats, is the major pollinator of *C. vinosa* and *C. longiflora,* which are restricted to the lowland wet forests around Golfo Dulce.

Thus, the bees probably are capable of pollinating a wider range of species than they have been observed to pollinate. The more varied the habitats of a plant species, the greater the number of bee species that will be encountered, hence, the less observed specificity.

Local environment: forest vs. open situations.—*Eulaema speciosa* and especially *Euglossa asarophora* tend to restrict their activity to the forest habitat, while *Eulaema nigrita, E. polychroma,* and to a lesser degree, *E. cingulata* commonly forage in open habitats or forest margins. A large population of *C. marantifolia* growing in an open pasture area near Monteverde was pollinated by both *E. cingulata* and *E. polychroma.* At Turrialba, the individuals of *C. marantifolia,* which were mostly in an open, cut-over area, were pollinated by *E. cingulata.* Those at the forest margin or within the forest were, however, also pollinated by *E. cingulata. Calathea vinosa* and *C. longiflora,* which are both pollinated by *Euglossa asarophora,* are deep-forest species, as are *C. silvicola* ssp. *silvicola* and *C. venusta.* The latter two, however, are pollinated by *Eulaema speciosa.* A relatively exposed garden population of *C. marantifolia* at Las Cruces was much more commonly visited by *Eulaema cingulata* than was the low-density forest population nearby, which was almost exclusively pollinated by *E. speciosa.* However, the difference in density between the two populations and some slight competition between *E. speciosa* and *E. cingulata* may better account for the observed difference in visitation to the two areas. For some reason, *E. cingulata* is infrequent on Barro Colorado Island and its role as the major pollinator of *C. marantifolia* and *C. latifolia* (in a semideciduous forest habitat) is taken over by *Euglossa imperialis.*

Influence of Tube Length on Pollinator Species

The relationship of corolla tube length to the tongue length of the insect visitors or bill length of birds has been noted often by pollination workers (Grant, 1949; Straw, 1956; Brian, 1957; Sprague, 1962; Percival, 1965).

Within the closed-flowered group of *Calathea* the tube lengths range from a mean of 42.9 mm. in *Calathea vinosa* to a mean of 23 mm. in *C. soconuscum* (table 5). For the wide-ranging species such as *C. marantifolia, C. macrosepala,* and *C. latifolia,*

TABLE 5
Corolla Tube Lengths of Closed-Flowered Calatheas

Plant Species	Tube Length X̄, mm.	Range	Standard Deviation	N
Calathea vinosa	42.9	41–45	0.127	14
C. foliosa	41.6	37–43	0.206	4
C. donnell-smithii	40.8	37–43	0.147	16
C. trichoneura	38.2	34–41	0.279	6
C. venusta	38.0	36–40	0.160	8
C. longiflora	37.0	35–41	0.195	10
C. warscewiczii	35.2	26–40	0.459	10
C. leucostachys	34.5	32–41	0.228	13
C. silvicola ssp. *tomentosa*	33.6	33–35	0.115	3
C. indecora	33.6	29–34	0.232	12
C. lasiophylla	30.4	28–33	0.159	10
C. macrosepala var. *reflexa*	29.4	27–30	0.105	8
C. silvicola ssp. silvicola	28.2	27–30	0.109	9
C. marantifolia	27.8	23–34	0.320	18
C. macrosepala var. *macrosepala*	27.0	23–29	0.196	11
C. latifolia	23.5	19–29	0.308	16
C. soconuscum	23.0	21–25	0.200	3

the mean of individual populations may vary markedly throughout the range. *Calathea latifolia* populations in Trinidad and southwestern Colombia have tubes averaging almost a centimeter shorter than those in Panamá.

The actual "reach" of the bee is determined by the combined lengths of the maxillae and prementum, and the glossa extending beyond the maxillae. The length of the glossa extended beyond the maxillae was 7.5 mm. in *Euglossa asarophora,* 6 mm. in *Eulaema speciosa,* 6 mm. in *Eulaema seabrai,* and 6 mm. in *Eulaema polychroma.* Thus, an approximate reach would be 35.5–39.5 mm. for *Euglossa asarophora,* 26.5–27.5 mm. for *Eulaema speciosa,* 25.5–26 mm. for *E. seabrai,* and 22–22.5 mm. for *E. polychroma.* These estimates are probably on the low side since *Euglossa asarophora* (estimated reach 35.5 mm.) is capable of robbing *C. donnell-smithii* with an average tube length of 40.8 mm. The robbing position employed by *E. asarophora* results in an "effective tube length" (tube length plus length of callose staminode) of *ca.* 48–50 mm.

The closed-flowered species with relatively small mean tube lengths (up to 28 mm.) are robbed by the longer-tongued *Eulaema meriana* and *E. bombiformis* but are pollinated by the shorter-tongued *E. cingulata, E. polychroma,* and *E. nigrita* (tables 2, 5, 6). *Eulaema seabrai,* though having a shorter proboscis than *E. meriana* or *E. bombiformis,* also robs these flowers (with mean tube lengths up to 28 mm.), but unlike *E. meriana* it feeds at the same level as the pollinators (that is, the actual tube length as opposed to the longer "effective tube length"). When visiting *Calathea donnell-smithii* (a longer-tubed species with a mean tube length of 40.8 mm.) on Cerro Jefe, *E. meriana* pollinated it. Although *E. speciosa* was prevalent on the Osa Peninsula, Costa Rica, it was never observed to visit *C. vinosa* (mean tube length 42.9 mm.) or *C. longiflora* (mean tube length 37 mm.), both of which were pollinated by *Euglossa asarophora*; but it regularly visited flowers of *C. inocephala*, a species with a shorter tube than either of the above. At Las Cruces, Costa Rica, the four clumps of *C. vinosa* were ignored by *Eulaema speciosa* until the number of *C. marantifolia* plants in bloom sharply declined, at which time they began to visit *C. vinosa* regularly. *Eulaema speciosa* effectively pollinated flowers with mean tube lengths of 27.8 mm. to 40.8 mm., and occasionally 42.9 mm., though when both a longer- and a shorter-tubed species were readily available in the area, it preferentially visited the shorter-tubed flowers. Brian remarked, in her work with *Bombus* (1957, pp. 77, 78), that the bees visited flowers having a tube length "considerably shorter than their tongue." This does not appear to be true in the case of the Euglossine pollinators.

DISCUSSION

There is a good correlation between corolla tube length and the ecology of the closed-flowered species of *Calathea*. In general, the species having the longest-tubed flowers (34.5–42.9 mm.—table 5) occupy the wettest habitats, are usually deep-forest species, and have a rather restricted distribution as regards moisture regimes. Those species having an intermediate tube length are found in wet to moist forest habitats, often in forest edge situations or occasionally in open areas. *Calathea marantifolia*, which has the broadest ecological range, occurring in wet to semideciduous forests and from sea level to 1400 m., has one of the widest ranges of tube lengths. Those species having the shortest tube lengths (23–27 mm.) occur in the driest habitats (for this group) and are rather shade intolerant, usually growing in open, often disturbed, areas.

A similar, but less striking, pattern prevails within the bee species that visit *Calathea*. Of the three species of *Euglossa*, *E. asarophora*, which has the longest proboscis, is a deep-forest bee, restricted to wetter habitats, while the other two species have broader ranges. *Euglossa imperialis* is commonly found in the semideciduous forest, as well as wet forest, and often forages in open areas. A similar trend is seen in *Eulaema*. Of the smaller-sized Eulaemas, *E. speciosa* is a wet-habitat forest species, while the three shortest-tongued species (table 6) range into drier habitats, occurring in the semideciduous forest zone and commonly foraging in open areas.

In addition to the habitat trend (wet to dry, deep forest to open areas) with decreasing tube length (table 5), there are similar trends in the number of bracts (few to numerous), the number of flowers produced per day (one or few to several [five to ten]), the population density (few scattered individuals to large stands), and seed size (large to small).

The amount and quality of nectar provided by a flower is related to the energy requirements of its pollinator or pollinators (Heinrich and Raven, 1972). When the amount of nectar per flower or the number of flowers per plant is limited, outcrossing is promoted since the pollinator is forced to visit several plants in order to meet its energy requirements (Heinrich and Raven, 1972). Plants that have several flowers

TABLE 6
Bee Tongue (Maxillae) Lengths

Bee Species	Maxillae Length, mm.
Euglossa asarophora	24–28
E. ignita	16–20
E. imperialis	16–18.5
Eulaema meriana	17–18.5
E. bombiformis	16–17
E. speciosa	16–17
E. seabrai	15–15.5
E. cingulata	13–14
E. polychroma	11.5–12
E. nigrita	11–12

per inflorescence and are closely spaced generally have less nectar per flower than plants with few flowers per inflorescence and with widely spaced individuals, since the pollinator must expend relatively more energy flying between the inflorescences. The two extremes in the closed-flowered Calatheas are *Calathea macrosepala* var. *reflexa* and *C. donnell-smithii. Calathea macrosepala* var. *reflexa,* which usually occurs in large stands and has several flowers per inflorescence per day, has small amounts of nectar (2.5 μl. was the largest of four samples), while *C. donnell-smithii,* which occurs in low-density populations and has one to two flowers per day, has an average of 25 μl. (N = 5). If the amount of nectar provided by an individual flower is too small relative to the energy required to gather it, a bee may not forage on the population (Heinrich, 1976). This would be the case where there were too few flowers in an inflorescence or the inflorescences were too far apart. This is seen in the pattern of visitation by the robber species to the closed-flowered Calatheas. The larger sized Eulaemas infrequently rob low density populations of closed-flowered species, while the frequency of their robbing increases dramatically on high-density plant populations, *e.g., C. latifolia* on Pipeline Road, Canal Zone, and *C. marantifolia* at Monteverde, Costa Rica. Since these robbers are larger bees than the pollinators, the amount of nectar provided by the above species when the plants are widely spaced may not provide sufficient caloric reward, but when the interplant distance is small, these plants become a viable resource.

In the closed-flowered Calatheas that produce several flowers per inflorescence per day, such as *Calathea latifolia* ($\bar{X}$ = 5.1, N = 45) and *C. marantifolia* ($\bar{X}$ = 4, N = 256), the majority of the pollinations probably result in geitonogamy. Probably about half of the pollinations in *C. donnell-smithii, C. longiflora,* and *C. venusta* result in cross-pollination since these three species have an average of less than two flowers per day. In general, the degree of outcrossing is higher for the deep-forest species, whereas geitonogamy prevails in the open-habitat species.

Harper *et al.* (1970, p. 345) noted, "In pairs of related species occupying open and closed communities it is normal to find the larger-seeded species in the more closed community. . . ." The distribution of seed size within the closed-flowered Calatheas reflects the relationship mentioned by Harper *et al.* The forest-dwelling species, such as *Calathea venusta,* produce larger seeds (9–10.5 mm. long and 5-6 mm. wide) than those of the open-habitat species, such as *C. macrosepala* var. *reflexa* (4.5-5 mm. long and 3.5–4 mm. wide).

DISTRIBUTION

LOCAL DISTRIBUTION

Plants of most species occur as scattered individuals or clumps within the forest and do not often form populations covering large areas. Since a single rhizome may often have several shoots, many of the clumps represent single individuals. They are not randomly dispersed but show a contagious (gregarious) distribution pattern and frequently are associated with a distinctive habitat, as along stream beds in the case of *Calathea longiflora* and *C. trichoneura,* or on the steep slopes above a stream, as in *C. foliosa.*

Though most species are found in moderate shade within the forest, the intensity of shade/sun is frequently an important factor in controlling their local distribution. The forest species that have the leaves clustered atop the stem, such as *Calathea donnell-smithii, C. foliosa,* and *C. venusta,* appear to have the greatest shade tolerance and are rarely found near the forest edge. *Calathea marantifolia* and *C. indecora* can occur scattered within the forest and also in large stands in open areas, the latter species often in areas of slightly soggy soil.

The species of open habitats, *Calathea macrosepala, C. soconuscum,* and *C. latifolia,* are shade intolerant, are more "weedy," and often form large, dense populations, especially in cut-over areas. They are nearly always found in open, fully exposed areas, often occurring close to the forest edge but not within the forest. In México, *C. macrosepala* has been found within a semideciduous forest, but the trees were widely spaced and the canopy was fairly open.

GEOGRAPHIC DISTRIBUTION

The closed-flowered Calatheas are predominantly Central American in distribution; the majority of species are found in Costa Rica and Panamá. Of the fifteen species, ten are found in Panamá and thirteen in Costa Rica; four of these are known only from Costa Rica. The species frequently are sympatric. Along the Atlantic coast of Panamá, *Calathea indecora, C. marantifolia, C. warscewiczii,* and *C. silvicola* ssp. *tomentosa* may occur within 100–200 m. of one another, although the local habitat occupied by each is usually different. On the Osa Peninsula of Costa Rica, *C. lasiophylla, C. vinosa, C. longiflora,* and *C. donnell-smithii* occur together, but again with noticeable differences in their local habitat. Of these four, *C. donnell-smithii* is the only species also occurring on the Atlantic slope of Costa Rica; the other three are apparently restricted to the Osa Peninsula and the area around Golfito by the dry Pacific slopes to the south and north and by the Cordillera de Talamanca to the east. The species most closely related to *C. vinosa* and *C. lasiophylla, C. indecora,* occurs along the Atlantic slope. *Calathea foliosa, C. indecora, C. warscewiczii, C. leucostachys,* and *C. venusta* occur on the Atlantic slope, the first only as far southeast as Bocas del Toro, Panamá, the last two also occurring near Portobelo, Panamá.

The range and distribution pattern of these species are largely influenced by the moisture regime. *Calathea leucostachys* and *C. venusta* are apparently absent from the lowlands around Almirante in Bocas del Toro, though they are present in the wetter forest areas near Siquirres, Costa Rica. *Calathea leucostachys* is also found in the premontane wet forests of San Ramón, Costa Rica, where there is a break in the cordillera that allows migration from (and to) the Atlantic lowlands. Those species with broader moisture tolerances, *C. warscewiczii, C. foliosa, C. indecora,* and *C. marantifolia,* are found around Almirante. The most drought-tolerant species, *C. macrosepala,* occurs in southern México and along the drier Pacific slope of Central America. The three species occurring in the deciduous to semideciduous forest habitats, *C. macrosepala, C. soconuscum,* and *C. latifolia,* have adapted to the long dry season. All three characteristically form large storage root-tubers near the end of the rainy season before going dormant. The seeds, shed during mid to late rainy season (August to November), remain dormant through the dry season (mid-December to March) and germinate at the beginning of the following rainy season.

The natural distribution of the closed-flowered Calatheas in Central America closely follows the distribution of the Euglossine bees. With the exception of *Euglossa flamea,* which has a more restricted range than does one of the plant species it pollinates, all of the other observed pollinator species range much farther south than any of the closed-flowered species (Moure, 1967; Robert Dressler, pers. comm.). The closed-flowered Calatheas only occur as far south as Venezuela (one species), northern and western Colombia (three species) and northern Ecuador (one species). Of the seven pollinator species, only two reach México, the other five ranging only as far north as Costa Rica. The closed-flowered Calatheas follow a similar pattern: three of the fifteen *Calathea* species occur in Guatemala or México, all the rest occur only as far north as Costa Rica (eleven species with one just getting into southern Nicaragua) or Panamá (one species).

The Calatheas and the Euglossines appear to have about the same altitude limit of around 1500 meters. At high elevations (1200–1500 m.) *Bombus* and *Thygater,* rather than *Euglossa,* were observed to be the main pollinators of the open-flowered *Calathea insignis.* Such a pollinator switch has not been observed for the closed-flowered species. However, a Colombian *Calathea* found at 2100 m. had red bracts and yellow flowers, colors more typical of hummingbird flowers. It is possible that the pollinators may impose a slight restriction on the altitude range of the closed-flowered Calatheas.

The distribution of the individual plant species within the closed-flowered group does not appear to be restricted by the distribution of any particular bee species, since the plants can be pollinated by any one of several Euglossines. Since non-Euglossine bees seem incapable of opening and pollinating the closed-flowered Calatheas and since these plants cannot set seed without the aid of a pollinator, the plants will be restricted to areas occupied by Euglossine bees, especially Eulaemas and Euglossas of the subgenus *Glossura.* The Euglossines are so widespread that it is difficult to say whether the closed-flowered Calatheas are actually restricted in their distribution by their pollinators without transplant and germination experiments in areas outside the range of the pollinators.

TAXONOMY

The type for the family, *Maranta arundinacea* L., was named by Plumier (1703): *Maranta arundinacea, Cannacori folio.* Koch (1857) was the first to recognize the Marantaceae as a family separate from the Cannaceae, stating:

> Meinerseits gebrauche ich die zuletzt erwähnte Benennung [Marantaceen] nur für die Scitamineen mit einfächrigem Staubbeutel, wo ein 1–3 samiger Fruchtknoten vorhanden und der Griffel an der Spitze und zwar meist hackenförmig gekrümmt ist, also für die Arten der Geschlechter *Maranta, Thalia, Phrynium* und *Calathea,* während ich mit dem Namen Cannaceen im engern Sinne die Arten mit mehr als 3-eiigen Fruchtknoten und nicht an der Spitze gekrümmten Griffel, also die des Geschlechtes *Canna,* verstehe.

Petersen (1889) recognized two major groups, the Phrynieae and the Maranteae, which Schumann later (1902) designated as tribes.

GENUS *CALATHEA*

Calathea G. F. W. Meyer, Primit. Fl. Essequeb. p. 6–7. 1818. Lectotype: *Maranta casupo* Jacquin chosen by S. Leman, Bull. Sci. Soc. Philom. Paris. 1820: 7. 1820.

Monostiche Koernicke Gartenfl. 7: 88. 1858. Type: *Phrynium coloratum* Hooker f., Bot. Mag. 4: *pl. 3010.* 1830.

Goeppertia Nees, Linnaea 6: 337. 1831. Lectotype (here chosen): *Maranta zebrina* Sims, Bot. Mag. 44: *pl. 1926.* 1817.

The genus *Goeppertia* was based on several heterogeneous elements. *Maranta zebrina* Sims is the only validly published element that fits the "capsula trisperma" of the protologue.

The genus *Calathea* is distinguished from the other tropical American genera by the presence of a 3-locular, 3-ovulate ovary. The name *Calathea* is derived from the Greek word kalathos meaning basket or vessel, because the species of this genus were thought to be used by the Indians (of northern South America) to weave baskets (Meyer, 1818). The portion of the plant used was not specified by Meyer. The various subgeneric classifications of *Calathea* are summarized in table 7. The closed-flowered Calatheas are in Schumann's (1902) series *Scapifoliae.*

SERIES *SCAPIFOLIAE*

The series *Scapifoliae,* characterized by an inflorescence of spirally arranged bracts subtended by a foliage leaf, as interpreted by Schumann (1902), included the following species: *Calathea humilis, C. rossii, C. pearcei, C. dicephala, C. sodiroi, C. pachystachya, C. longifolia, C. mandioccae, C. longibracteata, C. lindbergii, C. eichleri, C. violacea, C. cylindrica, C. allouia, C. macrosepala, C. leucostachys, C. picta,* and *C. warscewiczii* (table 8). *Calathea humilis* Spenc. Moore belongs in *Maranta* subgenus *Koernickea* and is not a *Calathea.* The position of *C. rossii* is still uncertain, but Koernicke (1862) states that it is "perhaps the same as *C. violacea* Lindl." *Calathea pearcei* and *C. dicephala* both have few bracts and small flowers accompanied by claviculate bracteoles, characters not present in the majority of the *Scapifoliae* or in the closed-flowered group. These species appear to belong in the subgenus *Microcephalum. Calathea sodiroi* and *C. pachystachya* are closely related to *C. altissima* and *C. elliptica* in the series *Nudiscapae.*

TABLE 7
Systems of Infrageneric Classification within *Calathea*

Koernicke (1858)
CALATHEA
- 1. Eucalathea
- 2. Anguste spicatae
- 3. Grandiflorae
- 4. Pseudophrynium*

Koernicke (1862)
CALATHEA
- 1. Eucalathea
- 2. Anguste vel brevissime spicatae
- 3. Grandiflorae
  - a. Folia cespitosa
    - 1. Spicae pedunculatae
    - 2. Spicae subsessiles
  - b. Folia solitaria
- 4. Pseudophrynium
  - a. Rotundifoliae
  - b. Pusillae
  - c. Bracteae ferrugineo-villosae vel lanata
  - d. Bracteae glabra vel plus minus pubescentes*

Bentham, in Bentham & Hooker (1883)
CALATHEA
- Section 1. Eucalathea
  - 2. Macropus
  - 3. Breviscapus
  - 4. Pseudophrynium*
  - 5. Microcephalum
  - 6. Monostiche

Petersen (1889)
CALATHEA
- Section 1. Eucalathea
  - 2. Macropus
  - 3. Breviscapus
  - 4. Pseudophrynium*
  - 5. Microcephalum
  - 6. Monostiche

Petersen (1890)
CALATHEA
- Section 7. Distichae
  - 6. Tubispathae
  - 1. Nudiscapae
    - a. Flores magni
    - b. Flores minores
  - 2. Scapifoliae*
  - 3. Rhizanthae
  - 4. Comosae
  - 5. Monostiche

Schumann (1902)
CALATHEA
- Subgenus 1. Eucalathea*
  - 2. Macropus
  - 3. Pseudophrynium
  - Series 3. Nudiscapae*
    - 1. Scapifoliae*
    - 4. Rhizanthe
    - 2. Comosae
    - 5. Polystachyae
- Subgenus 4. Microcephalum

*Includes the closed-flowered species of *Calathea*.

CLOSED-FLOWERED SPECIES WITHIN THE *SCAPIFOLIAE*

Calathea leucostachys, C. macrosepala, and *C. warscewiczii,* along with *C. latifolia, C. donnell-smithii,* and several species described since the publication of Schumann's monograph, form a natural grouping within the series *Scapifoliae.* The distinctive characteristic of these species is that the petals do not reflex at anthesis; the flowers remain closed as in bud and are forcibly opened by the pollinators. The corolla tube is fused throughout rather than having a slit on one side, and the capsule is rounded apically, not having a raised rim.

RELATIONSHIP OF CLOSED-FLOWERED SPECIES TO OTHER *SCAPIFOLIAE*

The closed-flowered group shares a similar habit, a similar leaf venation pattern, and spiral, imbricate bracts with the mainly open-flowered species in the series. Most

TABLE 8
Putative Species Relationships

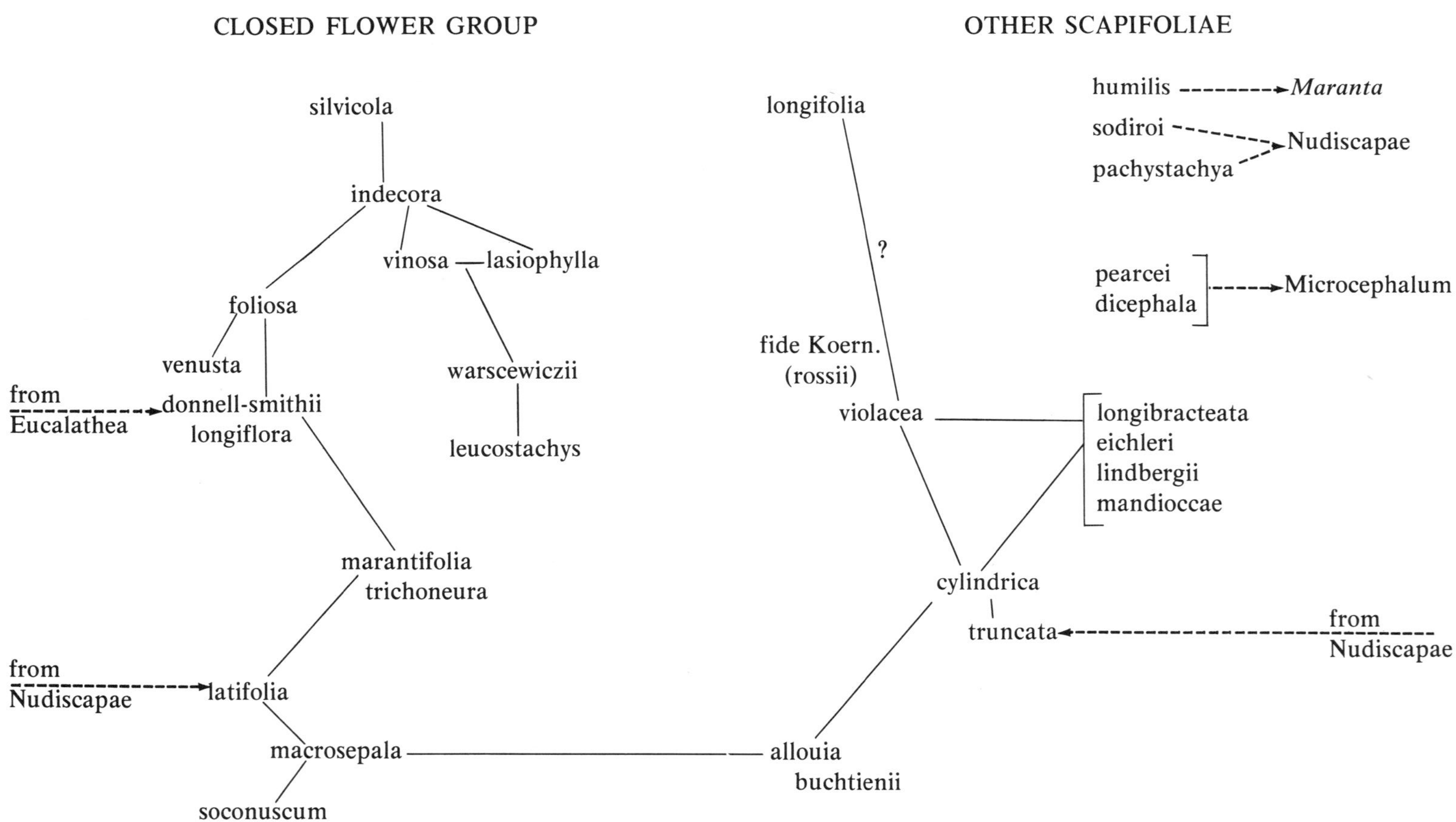

closely related to the closed-flowered taxa is *C. allouia,* a species of uncertain geographic origin presently found in the West Indies, the Meta Valley of Colombia, northern Ecuador, and the state of Pará, Brasil, mainly under cultivation. The shape and external morphology of the leaf are quite similar to those of *C. macrosepala,* and it also dies back to the rhizome seasonally, bearing numerous well-developed root-tubers. The flower, likewise, does not open spontaneously, but the sepals are thin and poorly developed (*ca.* 0.8 cm. long), the apex of the capsule not rounded, and the bracts much thinner and more herbaceous in texture than in *C. macrosepala. Calathea buchtienii* from Bolivia is nearly identical in inflorescence and bract shape to *C. allouia,* but has shorter and wider leaves and narrower vein spacing than *C. allouia. Calathea cylindrica* from coastal Brasil is more distantly related to the closed-flowered group. The petals of *C. cylindrica* do not reflex as in most open-flowered species, but move apart slightly at the apex. This gives a partially opened flower that is intermediate between an open-flower type and the closed flower of *C. allouia.*

SPECIES RELATIONSHIPS WITHIN THE CLOSED-FLOWERED GROUP

Based on morphological features, five species groups are recognized:

1. The drought-adapted species, *Calathea macrosepala, C. latifolia,* and *C. soconuscum.* Individuals of these three species bear well-developed root-tubers, die back to the rhizome during the dry season, and possess a pleated, relatively thin-textured leaf. *Calathea macrosepala* and *C. latifolia* share a similar bract morphology and texture and a similar capsule structure, the base of the capsule becoming swollen and mealy in appearance at the time of dehiscence. This character is not found in *C. soconuscum.* The flower color and morphology of *C. soconuscum* more closely resemble those of *C. macrosepala* than those of *C. latifolia,* the sepals of *C. latifolia* being more coriaceous and the flower usually larger.

2. The *Calathea marantifolia* and *C. trichoneura* alliance. Both species have coriaceous yellow bracts and sepals, a similar habit, and similar leaf texture and shape. *Calathea marantifolia* is widespread while *C. trichoneura* is restricted to higher elevations and moister habitats in Costa Rica and probably represents a relatively recent offshoot from the *C. marantifolia* lineage.

3. The *Calathea indecora* alliance, including *C. indecora, C. vinosa,* and *C. lasiophylla.* These species have numerous patent to recurved, densely tomentose bracts, the leaves are borne somewhat spread out along the stem, and a basal leaf is usually present.

Probably *Calathea silvicola* is closer to the *C. indecora* group than to the others, but its placement is uncertain. *Calathea silvicola* subsp. *tomentosa* is intermediate in habit between *C. indecora* and the subspecies *silvicola* (almost always bearing one or two basal leaves) as well as in inflorescence morphology (having densely pubescent, patent to strongly recurved bracts).

4. The *Calathea donnell-smithii* alliance, including *C. donnell-smithii, C. venusta, C. longiflora,* and *C. foliosa.* The first three share a similar inflorescence type, having a reduced number of glabrous or subglabrous bracts, as well as initially green bracts and sepals that turn purple with age. *Calathea foliosa* has yellowish, densely pubescent bracts. The characteristic habit with leaves clustered atop an elongated stem is

shared by *C. donnell-smithii, C. venusta,* and *C. foliosa. Calathea longiflora,* however, exhibits not only this habit, but also that of a basal leaf with the other leaves spread out along the stem.

While clearly allied with *Calathea donnell-smithii* and *C. venusta, C. foliosa* has densely tomentose, recurved bracts, a characteristic shared with the *C. indecora* group.

5. The *Calathea warscewiczii* and *C. leucostachys* alliance. Though more similar to each other in inflorescence and flower morphology than to other species, these two are probably less closely related than are the species in the other alliances.

KEY TO THE CLOSED-FLOWERED SPECIES OF *CALATHEA*

A. All leaves cauline, lowermost leaf arising above ground level; basal above-ground stem portion without adventitious roots . . . . P

AA. One or more leaves basal, lowermost leaf arising at or slightly below ground level, or if above ground, the stem rooting adventitiously at the nodes or through the lower leaf sheaths . . . . B

B. Upper leaf surface glabrous throughout except for midrib . . . . E

BB. Upper leaf surface minutely puberulent at base, villous throughout or appressed tomentose on the major veins C

C. Upper leaf surface and pulvinus villous throughout; apical portion of bract white; wet forest; southern Nicaragua to central Panamá . . . . 14. *C. leucostachys*

CC. Upper leaf surface minutely puberulent at base, at base and apex, or hairs restricted to the major lateral veins; pulvinus minutely puberulent (observed under magnification, not apparent to the naked eye) or glabrous, occasionally with a row of appressed hairs along the upper side; apical portion of bract yellow, green or green tinged with purple . . . . D

D. Major veins of upper leaf surface appressed tomentose, the intervein area and minor veins glabrous; leaf surface scalloped (tissue between major veins forming a smooth curve); bracts 5–13, bright yellow, turning blood red with age, upper half glabrous; premontane rain forest (*sensu* Holdridge) or cloud forest above 1000 m.; Costa Rica . . . . 5. *C. trichoneura*

DD. Major veins of upper leaf surface mostly glabrous, the intervein area pubescent (pubescence not restricted to major veins); leaf surface pleated (the tissue between each pair of veins major or minor planar); bracts 16–30, deep green, margin turning purple with age or yellowish with yellow green margin, pubescent throughout; wet forest; Atlantic slope of Costa Rica, Panamá, and Pacific slope of Colombia . . . . 9. *C. silvicola*

E. Upper surface of leaf uniformly green . . . . G

EE. Upper surface of leaf bearing a light green or yellowish green fishtail or jagged-edged pattern along the midrib . . . . F

F. Lower surface of leaf, petiole, and sheath dark red purple, leaf above bearing a yellowish green fishtail pattern which is 2-5 cm. across; lowland wet forest; Atlantic coast of Panamá and Costa Rica . . . . 15. *C. warscewiczii*

FF. Lower surface of leaf green, but occasionally bearing 2 light purple bands paralleling the midrib, petiole and sheath green, pattern narrow (usually less than 2 cm. across), the margin jagged; semideciduous forest; southern México and along Pacific slope from Guatemala to western Panamá . . . . 2. *C. macrosepala*

G. Lower surface of leaf dark red purple . . . . H

GG. Lower surface of leaf green I

H. Leaf scalloped, the blade shallowly curved between the raised major veins, semimetallic blue green above, lower surface densely covered with retrorsely barbed hairs (seen under magnification); inflorescence bracts straw-colored below with margins ivory; lowland wet forests surrounding Golfo Dulce, Costa Rica . . . . 6. *C. vinosa*

HH. Leaf pleated (tissue between two adjacent veins planar), the blade grass-green above, hairs straight when present on lower surface and usually restricted to the veins; inflorescence bracts green, margin green often tinged dark purple with age; Colombia and Panamá . . . . 9a. *C. silvicola* ssp. *silvicola*

I. Leaf pleated (leaf tissue between two adjacent veins planar) . . . . J

II. Leaf scalloped (tissue between raised major veins forming a smooth shallow curve); or leaf ridged, the major veins quite prominent, raised above the level of the intervein tissue, often only one minor vein between the major veins and these veins closely spaced (8–10 per cm., measurement taken halfway up the blade, *ca.* 2 cm. from the midrib) .... W

J. Bract margin patent to reflexed, tomentose to minutely puberulent (as seen under magnification, felt more than seen) .... N

JJ. Bract margin erect (in the same plane as the lower portion) or if slightly spreading to patent, both the margin and outer surface of the bract glabrous .... K

K. Sepals green to light green L

KK. Sepals cream white, lavender, or purple .... M

L. Bracts 9–53, outer surface minutely, uniformly puberulent (inconspicuous, though visible with the naked eye), green, if glabrous then the entire bract purple or green at base with the upper half purple; Panamá, Colombia, Venezuela, and Trinidad .... 3. *C. latifolia*

LL. Bracts 3–12, outer surface glabrous, green; southern México to northern Guatemala .... 1. *C. soconuscum*

M. In bud the 3 sepals meeting to form a blunt, rounded apex, the sepals purple, occasionally lavender; bract apex or entire bract deep purple; pulvinus round in cross section, not obviously laterally compressed, subglabrous to densely minute tomentose (as seen under magnification); 1–4 leaves cauline; Panamá, Colombia, Venezuela, and Trinidad .... 3. *C. latifolia*

MM. In bud the 3 sepals meeting to form an acute point at the apex, the sepals cream white, very rarely tinged purplish; bracts light green rarely tinged purplish (Costa Rica); pulvinus elliptical in cross section, noticeably laterally compressed, glabrous except for row of hairs above; 1 or occasionally 2 leaves cauline (arising on the stem above ground level); southern México to Coclé Province, Panamá .... 2a. *C. macrosepala* var. *macrosepala*

N. Shoots bearing 2 or fewer basal leaves; sepals grass green in flower; Panamá and Colombia .... 9a. *C. silvicola* ssp. *silvicola*

NN. Shoots bearing 3 or more basal leaves; sepals cream, lavender, or yellowish in flower .... O

O. Bracts light green throughout, light green tinged purplish apically, or light purple throughout; sepals cream white or lavender; Pacific slope, southeastern Costa Rica .... 2b. *C. macrosepala* var. *reflexa*

OO. Bracts light yellow below, reflexed margin yellowish green; sepals yellow or yellow tinged yellowish green at apex; Atlantic lowlands, Costa Rica, and western Panamá .... 9b. *C. silvicola* ssp. *tomentosa*

P. Bracts entirely green (grass green), green with dark purple margin, or entirely dark purple (depending on age); sepals green at anthesis, tinged with dark purple on maturing capsules ..... Q

PP. Bracts pale to bright yellow, greenish yellow, blood red, or cream, the recurved margin may be dark purple, blood red, or faintly tinged green; sepals cream to bright yellow at anthesis and may be tinged blood red or dark purple apically on maturing capsules .... T

Q. Petiole of subtending leaf at least 1.5 cm. long, usually 11.5 cm. long or greater; leaves pleated, 1–3 per shoot; bracts usually numerous, 12–30, occasionally more or less (10 in a Colombian specimen); Panamá and Colombia .... 9a. *C. silvicola* ssp. *silvicola*

QQ. Petiole absent in subtending leaf (rarely present, then less than 0.5 cm. long); leaves scalloped, 4–7 per shoot or leaves ridged, 2–4 per shoot; bracts few, 11 or less .... R

R. Leaf surface ridged, usually only 1 minor vein between the major veins, vein spacing close, 7–11 per cm.; 2–4 leaves per shoot, at least 1 of the lower leaves petiolate, 2 cm. long or longer .... 13. *C. longiflora*

RR. Leaf surface scalloped, usually 3 minor veins between the major veins, vein spacing wider, 4–7 per cm.; flowering shoot bearing 4–7 leaves, very rarely any of the leaves petiolate if petiolate, the petiole less than 0.5 cm. .... S

S. Leaves displayed in a spiral arrangement; bract margin patent to slightly reflexed; leaf narrowly elliptic, base acute; evergreen wet forest; Atlantic lowlands of Costa Rica and Panamá below 200 m. .... 11. *C. venusta*

SS. Leaves distichous, arranged in a single plane; bract margin erect; in young, inflorescences clasping the base of the bract above; leaf ovate to elliptic, base rounded to subtruncate; generally premontane wet to cloud forest; Costa Rica and Panamá above 500 m., except near sea level on the Osa Peninsula, Costa Rica .... 12. *C. donnell-smithii*

T. Upper leaf surface appressed-tomentose along the major veins, intervein areas glabrous; premontane rain forest or cloud forest above 1000 m.; Costa Rica .... 5. *C. trichoneura*

TT. Upper leaf surface entirely glabrous except for puberulent midrib .... U

U. Bracts densely velvety-tomentose abaxially, the apical margin (adaxial surface) recurved to revolute, velvety-tomentose (visible with the naked eye) V

UU. Bracts appressed-tomentose basally, glabrous to minutely puberulent (seen under magnification) above, the apical margin not recurved or patent though occasionally slightly spreading, the adaxial surface glabrous to minutely puberulent (seen under magnification); Guatemala to Ecuador. .... 4. *C. marantifolia*

V. Lower surface of leaf glabrous; leaves tightly clustered above an elongated stem internode, this internode (between the innermost cataphyll and first leaf) comprising more than 1/2 the height of the plant, the internode between the subtending and next lower leaf less than 4 cm. long; bract margins turning purple with age; Atlantic slope of Costa Rica and western Panamá....... .... 10. *C. foliosa*

VV. Lower surface of leaf densely velvety-pilose; leaves spaced out along the stem length; basal stem internode comprising less than 1/4 the height of the plant, the internode between the subtending and next lower leaf 15 cm. long or greater; no color change in bract margins with age; evergreen wet forest around Golfo Dulce, Costa Rica. .... 7. *C. lasiophylla*

W. Leaf ridged, veins closely spaced (7–11 per cm.); bracts grass green turning dark purple with age, abaxial basal portion minutely puberulent, upper portion of the bract and the patent to reflexed apical margin glabrous; sepals green at anthesis turning dark purple apically on maturing capsule; corolla tube 3.6–4.1 cm. long; evergreen wet forest around Golfo Dulce, Costa Rica. .... 13. *C. longiflora*

WW. Leaf scalloped, veins widely spaced (5–7 per cm.); bracts pale to bright yellow, pale yellowish green or straw-colored at base with greenish white margin, the margins densely tomentose when patent or reflexed; sepals yellow, cream, or yellow tinged yellow green apically; corolla tube less than 3.4 cm. long. X

X. Bract margin patent to reflexed, margin and outer surface of bract uniformly densely tomentose; basal portion of bract pale straw color, margin tinged faint green or pale yellowish green........ Y

XX. Bract margin erect (in the same plane as lower portion), basal portion of bract densely appressed-tomentose thinning apically, the upper half glabrous to minutely puberulent, the pubescence at the base denser and longer than that in the apical portion, the inner surface of the bract glabrous to minutely puberulent (seen under magnification); bract bright to pale yellow, occasionally greenish yellow; widespread, Guatemala to Ecuador................. 4. *C. marantifolia*

Y. Lower surface of leaf densely velvety-pilose throughout, seen with the naked eye; bracts equal to or greater than 3.3 cm. high; bicarinate prophyll at least 0.6 cm. shorter than the bract; evergreen wet forest around Golfo Dulce, Costa Rica. .... 7. *C. lasiophylla*

YY. Lower surface of leaf glabrous or puberulent (seen under magnification) only along the veins, felt rather than seen; bracts equal to or less than 2.7 cm. high; bicarinate prophyll usually equal to or exceeding the bract in length; lowland wet to premontane wet forest; Atlantic slope of Costa Rica and western Panamá. .... 8. *C. indecora*

1. *Calathea soconuscum* Matuda

(Plate 4, Figure 5)

Calathea soconuscum Matuda, Ann. Inst. Biol. México, 21(2): 327, 328. 1950. Type: Orilla del Río Doña Maria, en un lugar "Caricanto," bosque bajo, arenoso, 100 m., Acacoyagua, Chiapas, México, 26 July 1950, *K. Nakamura 27* (MEXU 177035!, ex MATUDA Herb. 19345; isotypes MEXU 89551!, s.n.!).

Caulescent herb 0.66–1.32 m. high. Each shoot bearing 2–4 leaves, 1–3 basal, 1 or less commonly 2, cauline. Roots bearing numerous root-tubers, 1.5–2.5 cm. long and 1–3 cm. wide (measurements *fide* Matuda). The plant dies back to the rhizome with attendant "tubers" during the dry season, resuming growth and flowering during the rainy season. *Cataphylls* 2–3, narrowly ovate, acute, green, often streaked with purple, innermost one 23–68 cm. long. *Leaf blade* herbaceous, major veins raised, leaf with pleated appearance, ovate, obtuse to obtuse with an acuminate point, rounded to subtruncate at base, subtending leaf 21–41 cm. long and 15.3–21.5 cm. wide, others 25–52 cm. long and 13.5–26 cm. wide. Leaf surface

TABLE 9

Distribution of Certain Features in the Closed-Flowered Species of the *Scapifoliae*

X = present throughout species
\- = present in part of species (not unusual)
o = occasionally present
r = very rare
X* = faint greenish tinge
X⁺ = may be > 0.5 cm but < 3 cm

	soconuscum	*macrosepala* var. *macrosepala*	var. *reflexa*	*latifolia*	*marantifolia*	*trichoneura*	*vinosa*	*lasiophylla*	*indecora*	*silvicola* ssp. *silvicola*	ssp. *tomentosa*	*foliosa*	*venusta*	*donnell-smithii*	*longiflora*	*leucostachys*	*warscewiczii*
leaf purple below							X			r						-	X
leaf green below or green with purple bands	X	X	X	X	X	X		X	X	-	X	X	X	X	X	-	
sepals initially cream or white		X	-		r		X*	X*								X	X
sepals initially yellow					-	X			X		X	X					
sepals initially green	X		-	o						X			X	X	X		
sepals initially purple or tinged purple			o	-	r												
bracts initially cream or white							X	X	X*							X	X
bracts initially yellow					X	X					X	X*					
bracts initially green	X	X	-	r						X			X	X	X		
bracts initially purple or tinged purple			-	-													
bracts and sepals changing to red with age						X											
bracts and sepals changing to purple with age										X	X	X	X	X	X	X	X
bracts and sepals having no marked color change	X	X	X	X	X		X	X	X								
apex of bract patent to recurved	o		X			X	X	X	X	X	X	X			X	X	
apex of bract erect	-	X		X	X									X			
apical bract margin glabrous adaxially	X	X		X	-	X						X	X	X			
bicarinate prophyll glabrous	X	X	-	-	-	X						X	X	X			
sheath auriculate														X			X
petiole absent or less than 0.5 cm. in all leaves						r							X	X	X⁺	X⁺	
petiole present in at least 1 leaf, over 3 cm. long	X	X	X	X	X	-	X	X	X	X	X	X					X
pulvinus pubescent			-	-							X					X	X
at least 1 leaf always basal	X	X	X	X			X		X		X						X
leaves usually basal and cauline	X	X	X	X	X	X	X	X	X	X	X				X	X	X
leaves always entirely cauline												X	X	X			
leaf ridged															X		
leaf scalloped					X	X	X	X	X			X	X	X		X	X
leaf pleated	X	X	X	X						X	X						
plant evergreen					X	X	X	X	X	X	X	X	X	X	X	X	X
plant dying back to the rhizome	X	X	X	X													

above grass green, semiglittering, glabrous, midrib green, slightly lighter than blade, bearing a row of hairs centrally; lower surface pale grey green, glabrous to subglabrous, midrib light olive green. *Pulvinus* yellow green to pale olive green, glabrous except for row of hairs above, round in cross section, 2.5–7.5 cm. long, acuminately extended along the underside of the petiole. *Petiole* green, sparsely pubescent, round in cross section, not bearing a groove, 2.6–27 cm. long. *Leaf sheath* papyraceous to herbaceous at margins, grass green occasionally striped purple at the base, puberulent along the center back, margins subglabrous, 9–21.5 cm. long in subtending leaf; others 38–85 cm. long. *Stem* green.

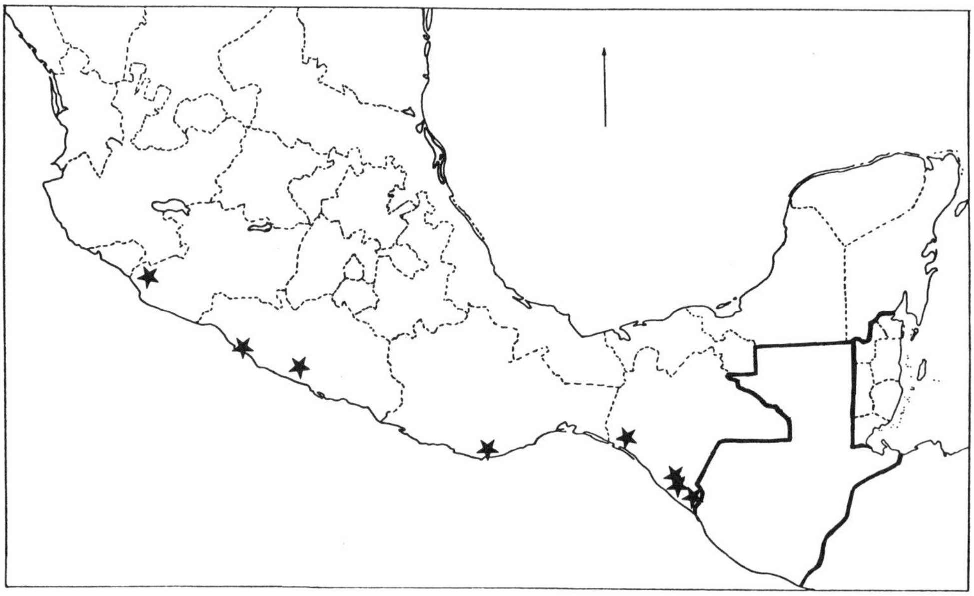

Fig. 5. Distribution of *Calathea soconuscum* in México.

Inflorescence terminal subtended by a foliage leaf, obconical, 6–8.5 cm. high and 5–6 cm. wide. *Peduncle* green, glabrous, 8.7–23 cm. long. *Bracts* 3–12, spirally arranged, semicoriaceous, base imbricate, apical 1/2 somewhat spreading, transverse-elliptical to depressed-ovate, apex rounded to obtuse, margin usually erect, occasionally patent to slightly recurved, 2.2–2.8 cm. high and 3.6–4.4 cm. wide, each subtending at least 5 flower pairs. Outer surface shiny grass green, glabrous; inner surface shiny green, glabrous. *Bicarinate prophyll* membranaceous, obovate-elliptic, obtuse, translucent light green, glabrous, (2.1)2.2–2.7 cm. high and (1.1)1.5–1.8 cm. wide. *Secondary bracts* membranaceous, broadly oblong to elliptic, apex rounded, translucent green above, basal 1/4 pale green to whitish, glabrous, (1.4)1.7–2.3 cm. high and (1.4)1.6–2.05 cm. wide. *Bracteoles* membranaceous, usually 2 per flower pair, narrowly ovate, frequently one channeled and the other carinate, or both carinate, faint greenish, glabrous 1.2–1.9 cm. high and 0.4–0.5 cm. wide.

Flowers never open spontaneously. *Sepals* semispongy at base and centrally, margins herbaceous, channeled, narrowly obovate, apex obtuse to subrounded, shiny, lower half white, green above, glabrous, (1.6)1.8–2.9 cm. high and (0.5)0.6–0.95 cm. wide. *Corolla tube* cream, glabrous, (2.1)2.3–2.5 cm. long; corolla lobes unequal, elliptic to obovate-elliptic, apex obtuse to 90°, golden orangish yellow, tips tinged greenish, (1.2)1.4–1.6 cm. long and (0.55)0.75–1.15 cm. wide. *Outer staminode* obovate to obtriangular, rounded to emarginate, apical portion dished, pale cream, 1–1.2 cm. long and 0.75–0.9 cm. wide. *Callose staminode* rectangular, apex rounded on one side, bidentate on the other, pale orangish, apical margin often tinged red purple, 0.9–1 cm. high. *Cucullate staminode* cream white at the tip, mid and basal portion light orange, 0.9–1 cm. long and *ca.* 0.45 cm. wide. *Filament* with lateral petaloid appendage, adnate to style and callose staminode, upper 1 mm. free; anther free, 3 mm. long. Style and stigma golden. *Ovary* cream, glabrous, 2–3 mm. high. *Capsule* obovoid, 0.9–1 cm. high and 0.65–0.7 cm. wide, crowned by a persistent, live, expanded calyx. *Seeds* usually 3, trigonous, rugose on outer surface, slate grey, 5.5–6 mm. high and 3.5–4 mm. in diameter, bearing a basal white aril 2–3 mm. high.

Representative specimens seen: MÉXICO. Chiapas: Around the microwave tower and stream below it, Quetzalapa Est. Microondas, *ca.* 14 mi. S.E. of Tonalá on Highway 200, Munic. Tonalá, 1300 ft., 11 Aug. 1972, *H. Kennedy, G. L. Webster, S. Lynch 1462* (F, MEXU, US); Esperanza, Escuintla, 12 Aug. 1947, *E. Matuda* (Herb.) *16074* (MO); Hidalgo, Escuintla, in open forest, 21 June 1947, *E. Matuda* (Herb.) *16592* (F, MEXU), cited in original description; in lowland thickets, Islamapa, Huehuetan, 27 June 1948,

E. Matuda (Herb.) *18012* (MEXU). Oaxaca: Cafetal Concordia (Cerro Espino), Aug. 1917, *B. P. Reko 3305* (US); in Barbaseti, humids solicalcari pr. Rancho Chacalapa inter la Galera et Pochutla, Oct. 1842, *Liebmann 14706* (C). Guerrero: Zihuatanejo, 30 m., 7 Aug. 1898, *E. Langlasse 275* (GH, K, P, US). Michoácan: Villa Victoria, Distr. Coalcoman, forest, 500 m., 9 July 1939, *G. B. Hinton 13896* (GH, NY, US).

Flowering: *Calathea soconuscum* flowers in early rainy season, June to August. Plants collected in mid-August were no longer flowering but had well-developed capsules. Peak flowering occurs earlier (about a month or two) in *Calathea soconuscum* than for the sympatric *C. macrosepala*. When the latter is in good bloom, *C. soconuscum* is nearly finished blooming.

Distribution and habitat: *Calathea soconuscum* occurs from sea level to about 400 m. along the Pacific lowlands of southern México, from northern Michoacán south and east to southern Chiapas. This species occurs in deciduous forest usually in the moister areas along streamsides. It may also occur in "thickets" or more or less open secondary growth or sparsely wooded areas near a stream. Although it has not as yet been reported, it would be expected in northern Guatemala.

Common name: *Macuse, makuce*; the label on the type specimen reads *Macuse simarron*. These names are used in Chiapas, México.

Variation: The leaf shape of *Calathea soconuscum* from Michoacán (*Hinton 13896*) is elliptic to broadly elliptic which is much more rounded than that of the leaves of the Chiapan populations.

Relationship: The habit of *Calathea soconuscum* is similar to that of *Calathea marantifolia* and *C. latifolia*. The leaf is thin, herbaceous, and pleated in appearance, characteristics shared with both *C. latifolia* and *C. macrosepala*; however, the capsule does not swell at the base at the time of dehiscence as it does in the latter two. The sepals tend to be somewhat pointed in bud, similar to *C. macrosepala*. The differences in habit and capsule between *C. macrosepala* and *C. soconuscum* suggest it may be closer to *C. marantifolia* and the similarities with *C. macrosepala* due in part to its being adapted to a drier habitat.

2. *Calathea macrosepala* K. Schumann

This is a wide-ranging species, known from Chiapas, México to western Panamá, and is most abundant in the northwestern part of the range. A second variety is recognized that is restricted to the Pacific slope in southern Costa Rica.

Key to the varieties of *Calathea macrosepala*

a. Margin of bract erect, adaxial apical margin glabrous .... var. *macrosepala*
aa. Margin of bract recurved to revolute, recurved (adaxial) margin densely minute puberulent .... var. *reflexa*

2a. *Calathea macrosepala* var. *macrosepala*

(Plate 5, Figures 6, 7)

Calathea macrosepala K. Schumann, Pflanzenreich IV.48: 84. 1902. Lectotype (here chosen): Guatemala, Dept. Santa Rosa, La Vega, 1600 m., Aug. 1893, *Heyde* and *Lux 6256* (K!; isolectotypes BM!, GH!, US!).

Rhizomatous herb 0.8–1.8 (2.2) m. high. Shoots bearing 4–8 leaves, 3–7 basal, 1–2 cauline. Plants dying back to the rhizome during the dry season, putting up a new shoot and flowering in the rainy season. Roots bearing root-tubers up to 3 cm. wide and 6 cm. long in adult plants. *Cataphylls* herbaceous, narrowly ovate, apex acute to obtuse, mucronulate, green, scattered tomentose along back, more dense to appressed tomentose at base, innermost cataphyll 17–30 cm. long. *Leaf blade* herbaceous, narrowly oblong-elliptic to narrowly ovate, apex abruptly acuminate to abruptly attenuate acuminate, base unequal, obtuse to rounded, subtending leaf (33)37–90 cm. long and (14)19.5–33.5 cm. wide; others, (34)46.5–82 cm. long and (14)23.5–34.5 cm. wide. The lowermost leaf relatively small, the subtending or penultimate leaf the largest. Leaf blade pleated, major lateral veins raised, upper surface slightly glittering, nearly always bearing a narrow feathery pale yellow green pattern along the midrib, especially noticeable in young plants and in lower leaves, glabrous, except for the row of hairs along the midrib; lower surface pale greyish to slight silvery green, commonly with 2 broad purple bands, puberulent to subglabrous, midrib yellow green. *Pulvinus* elliptical in cross section, yellow green to light olive green, at junction with

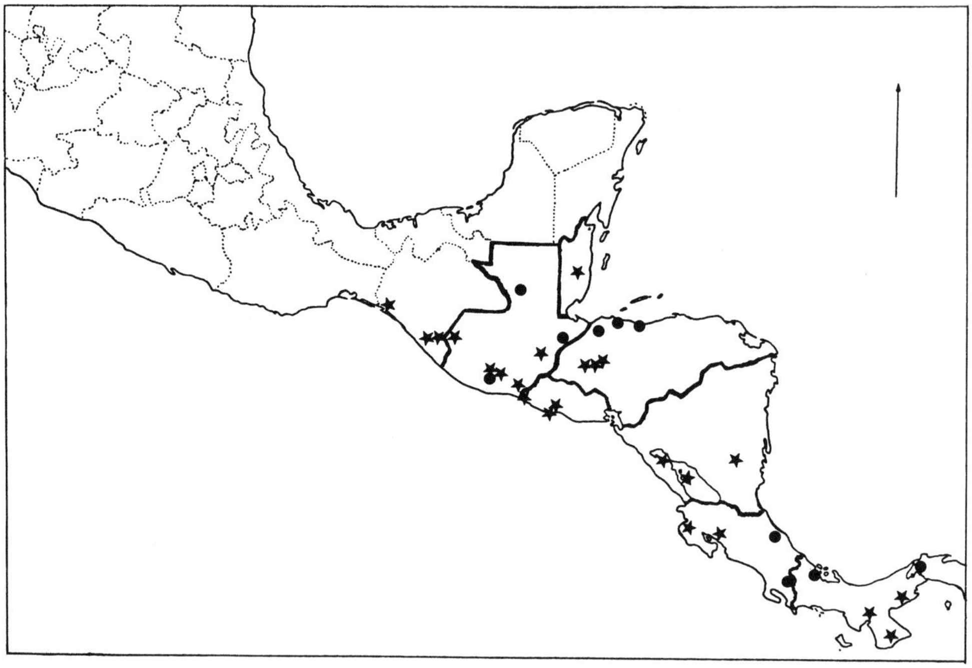

Fig. 6. Distribution of *Calathea marantifolia* (circles) and *C. macrosepala* var. *macrosepala* (stars) in Central America.

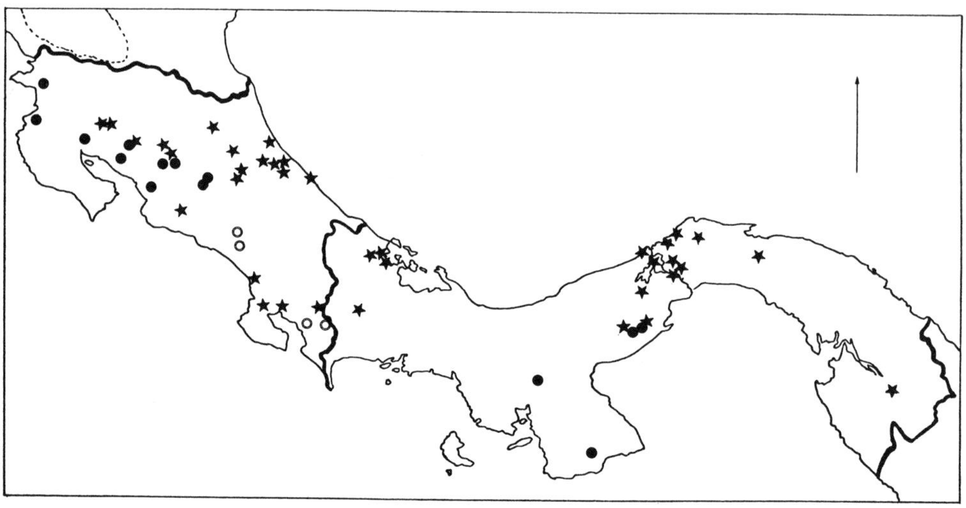

Fig. 7. Distribution of *Calathea marantifolia* (stars), *C. macrosepala* var. *macrosepala* (solid circles), and *C. macrosepala* var. *reflexa* (open circles) in Costa Rica and Panamá.

petiole acuminately extended along the upper side of the petiole and marked by a greenish yellow ring, glabrous except for row of stiff hairs above, (0.6)1.1–5.5 cm. long in subtending leaf; others (0.7)1.5–4 cm. long. *Petiole* often absent, when present bearing a narrow groove *ca.* 1 mm. wide and 1 mm. deep along the upper surface, grass green, glabrous, 0–13.5(17.5) cm. long. *Leaf sheath* not auriculate, herbaceous, green, occasionally with purple striping where covered by another or at ground level, lighter yellow green just above the nodal area, apical margin of the leaf sheath densely tomentose; below, tomentose along margin to subglabrous toward base, central portion subglabrous, densely appressed pilose in basal cm., (9.5)14.5–34.5 cm. long in subtending leaves; others 31.5–115 cm. long. *Stem* grass green, puberulent just below nodes, the rest subglabrous to glabrous.

Inflorescence terminal, subtended by a foliage leaf, imbricate, ovoid to ellipsoid, 6–11 cm. high and 4–5.5 cm. in diameter. *Peduncle* green, puberulent above, subglabrous below, 6.5–21 cm. long. *Bracts* 26–41, spiral, imbricate but not tightly so, subcoriaceous, depressed ovate-trullate to broadly trullate, apex obtuse, 2.4–3.2 cm. high and 3–5.5 cm. wide, the basal-most bracts wider in proportion than upper ones. Outer surface light green, minutely puberulent, more densely so toward base of the bract; within, shiny light green, glabrous. Each bract may subtend at least 7 flower pairs. *Bicarinate prophyll* membranaceous, circular to transverse broadly elliptic, apex rounded, translucent light green, glabrous, infrequently the apical margin puberulent, (2.1)2.2–2.8 cm. high and (1.9)2.2–3.2 cm. wide. *Secondary bracts* membranaceous, broadly oblong to transversely oblong, apex truncate to subretuse, translucent light green, glabrous, 2.2–2.45 cm. high and 1.7–2.6 cm. wide. *Bracteoles* subtending flowers almost always 2 per flower pair, membranaceous, glabrous, 1.8–2.2 cm. high and 0.4–0.6 cm. wide.

Flowers never opening spontaneously. *Sepals* subherbaceous, channeled, narrowly elliptic to narrowly obovate, apex obtuse, in bud the 3 sepals meeting to form an acute point, shiny, white at base, faint cream above, glabrous, 1.8–2.8 cm. long and 0.5–0.8 cm. wide. *Corolla tube* cream at base becoming light orangish above, 2.3–2.9 cm. long; corolla lobes unequal, elliptic to broadly elliptic or obovate-elliptic, apex obtuse, golden orange to golden yellow, 1.4–1.7 cm. long and 0.9–1.1 cm. wide. *Outer staminode* unguiculate to obovate, apex rounded, irregular, slightly cupped at tip, cream to white, 1–1.2 cm. long and 0.4–0.7 cm. wide. *Callose staminode* rectangular, truncate with a short acumen to one side or bidentate, golden orange, the apical margin dark red purple, 1–1.2(1.3) cm. high. *Cucullate staminode* dark golden orange basally, the hooded portion pale orange, 0.85–1.1 cm. long and (0.45)0.5–0.6(0.7) cm. wide. *Filament* with reduced lateral petaloid appendage *ca.* 1.5 mm. wide, pale orangish; anther free, 3–3.5 mm. long. Style and stigma deep golden orange. *Ovary* white, glabrous 2.5–3 mm. high. *Capsule* obovoid, rounded at the apex, smooth, semishiny, cream to yellowish, 1.2–1.3 cm. high and 0.7–0.8 cm. in diameter, abscissing from the pedicle at the time of dehiscence, the tissue at the base is swollen and mealy at the time of abscission. The capsule is crowned by a persistent enlarged calyx. *Seeds* usually 3 per capsule, trigonous, rugose on outer surface, slate grey, 5.5–6 mm. high and 4–4.5 mm. in diameter, bearing a basal aril 3–3.5 mm. high.

Representative specimens seen: PANAMÁ. Coclé: km. 21 on road to El Valle, 13 Aug. 1972, *A. Gentry 5615* (MO); western slope and summit of Cerro Valle Chiquito, 700–800 m., 25 July 1935, *R. J. Seibert 647* (F, GH, K, MO, NY). Los Santos: from 1 mi. S. to 10 mi. N. of Tonosí, 14 July 1967, *J. A. Duke 12479* (MO); dry deciduous woods north of the village of Guaniquito about 10 mi. N. of Tonosí along Río Tonosí, 100–200 m., 17 July 1970, *J. L. Luteyn* and *R. B. Foster 1365* (DUKE). Veraguas: Santigo, 2 mi. W. of Santiago on Pan Am Highway, 6 Aug. 1967, *J. D. Dwyer et al. 7560* (MO).

COSTA RICA. Puntarenas: Cascajal (25 km. E.S.E. of Puntarenas) Valley of Machuca River, open second growth woods on steep slopes, 5 July 1949, 30–100 m., *R. W. Holm* and *H. H. Iltis 253* (A, P); Lindora lighthouse, near Monteverde, *ca.* 900 m., 29 Sept. 1970, *H. Kennedy 567* (CR, F, MO, US); shaded pasture, 1 km. N. of the Inter-Amer. Highway, along the road to Monteverde, sea level, 11 July 1972, *J. L. Luteyn 3393* (DUKE). San José: secondary growth along Río Virilla, 4 km. N. of the center of San José, 1100 m., 22 Sept. 1971, *R. W. Lent 2164* (F); fossés á La Verbena prés Alajuelita, 1000 m., Aug. 1894, *H. Pittier 8832* (BR, US); buissons au pont du Virilla prés San Juan, 1100 m., 27 Sept. 1895, *A. O. Tonduz 9775* (BR, CR, US), US 471939 reads 1200 m., July 1896. Guanacaste: pastureland and deciduous forest now in full leaf in the areas of Bahia El Coco, Bahia Playa Hermosa and Sardinal, 0–150 m., 10° 32′ N. x 85° 40′ W., 29–31 July 1971, *W. C. Burger* and *M. Burger 7733* (F); Parque Nac. Sta. Rosa, 22 July 1971, *R. Callaway 171* (CR); along road to gallery forest from field station at Finca Taboga and in gallery forest, secondary growth, ca. 12 m., 10° 19′ N. x 85° 09′ W., 8 Aug. 1968, *H.*

Kennedy 63 (US); Finca Taboga, 6 mi. S., 6 mi. W. of Cañas, along agricultural service road, wet lowland, 19 July 1968, *R. G. Koch 4880* (MO).

NICARAGUA. Rivas: Isla d'Ometépé bois humides, 100 m., Oct. 1869, *P. Levy 1176* (B, P). Managua: south of Managua, Las Nubes and vic., 850 m., 27 June 1923, *W. R. Maxon, A. D. Harvey, A. T. Valentine 7449* (US); 800-900 m., 28 June 1923, *W. R. Maxon et al. 7495* (US).

EL SALVADOR. San Salvador: San Salvador, 1922, *S. Calderón s.n.* (NY, US); San Salvador, 657 m., July-Aug., *Labr. Quimico 1* (US). La Libertad: 9 km. E. of La Libertad, very shaded, damp volcanic humus, 200 m., 15 Aug. 1938, *O. B. Horton 8650* (K, US). Ahuachapán: 1923, *S. A. Padilla 588* (US).

HONDURAS. Cortés: matorrales muy humedos entre Agua Azul y Pito Solo, area Lago Yojoa, 600 m., 26 Aug. 1955, *A. Molina R. 5540* (F); montaña al Jaral, 28 Aug. 1955, 600 m., *A. Molina R. 5687* (F). Santa Barbara: cut-over pine-oak forest, between Ceguaca Abajo and Concepción del Sur, 21 Aug. 1968, *A. Molina R. 21961* (F).

BELICE. Dept. unknown: Jacinto Hills, secondary forest, shade, 100 ft., 8 Nov. 1933, *W. A. Schipp 8-618* (F); 300 ft., 1 Mar. 1934, *W. A. Schipp 8-919* (F).

GUATEMALA. Zacapá: Sierra de las Minas, pine-covered slopes, along trail between Río Hondo and summit of Mt. at Finca Alejandría, 1000-1500 m. in ravine, 11 Oct. 1939, *J. A. Steyermark 29590* (F). Escuintla: Chimaltenango, San Andrés Osuña, 19 Aug. 1937, *J. R. Johnson 992* (F). Santa Rosa: Jumaytepeque, 6000 pp., Aug. 1892, *Heyde et Lux 3872* (GH, K, MO, US); *Heyde et Lux 3876* (K); hills E. of Cuilapa, along the stream supplying the city water, 900-950 m., 24 Nov. 1940, *P. C. Standley 78199* (F). Huehuetenango: canyon, tributary to Río Trapichillo, between Democracia and canyon of Chamushú, 1000-1100 m., 24 Aug. 1942, *J. A. Steyermark 51237* (F, MO). Dept. unknown: Secanquim trail to Sepacuite, 28 Nov. 1904, *G. P. Goll 93* (US); bank of Río Brevo, Finca Mocá, Guatalon, 20 Aug. 1936, *W. R. Hatch and C. L. Wilson 378* (F).

MÉXICO. Chiapas: around the microwave tower and stream below it, Quetzalapa Est. Microondas, *ca.* 14 mi. S.E. of Tonalá on Highway 200, Munic. of Tonalá, 1300 ft., 11 Aug. 1972, *H. Kennedy, G. L. Webster, S. Lynch 1465* (F); along riverbank in secondary forest, near Belisario on the road from Huixtla to Motozintla, ca. 3000 ft., 13 Aug. 1972, *H. Kennedy, G. L. Webster, S. Lynch 1473* (MEXU, US); Acacoyagua, orilla del Río Doña Maria en lugar "Caricanto," bosque bajo, arenoso, 100 m., 26 July 1950, *K. Nakamura 27* (MEXU); Calaluta, Acacoyagua, 150 m., 28 July 1950, *K. Nakamura 28* (MEXU); en bosque alta humeda, Calaluta, Acacoyagua, 150 m., 26 July 1950, *E. Matuda 19346* (MEXU); Escuintla, Esperanza, 2 Nov. 1941, *E. Matuda 16074* (F, MEXU, MICH); Vega, Santa Maria, Río Mixcun, cerca Guatimoc, bosque secondaria, 4 Dec. 1941, *F. Miranda 1807* (MEXU). Veracruz: San Lorenzo, Tenochtitlan, loma de los gringos, 28 Nov. 1967, *J. Chavelas P., M. Esparza, L. Aceves s.n.* (MEXU).

Flowering: *Calathea macrosepala* var. *macrosepala* flowers during the mid and late rainy season, July to November; in Belice, it flowers as early as March.

Pollinator: In August, 1972, *Calathea macrosepala* var. *macrosepala* was observed to be pollinated by *Eulaema cingulata* (Fabricius) near Tonalá, Chiapas, México.

Distribution and habitat: *Calathea macrosepala* var. *macrosepala* has a wide distribution, from the state of Veracruz in southern México, along the dry, Pacific escarpment of Central America south and east to Coclé Province in Panamá. In Chiapas, México, it occurs in gallery forests along rivers, and in seasonal deciduous to semideciduous forest on the lower slopes of the Cordillera de Tilerán, Costa Rica, to premontane wet forest below Monteverde, Costa Rica. At the Monteverde locality, the plants were not in the forest but on an exposed slope. In Panamá it occurs in seasonal deciduous forest. It most frequently occurs on moderate to steep slopes. In northern Puntarenas Province, Costa Rica, it is especially prevalent on cut-over banks above a stream, and on the steep high banks of the Río Virilla, Costa Rica, it is frequently in open exposed areas. It occurs in a semideciduous forest in Chiapas, in which the trees are scattered and provide slight shade.

Common name: It is known as *chufle* in El Salvador and as *macuse* in Chiapas.

Variation: For one with such a wide distribution, *Calathea macrosepala* var. *macrosepala* is remarkably uniform. The bract apex tends to be broadly obtuse in the Chiapan individuals, and is more pointed in the southern populations. In addition, the pubescence is denser at the base of the bract in the Chiapan plants, whereas it is more uniform throughout the bract in Costa Rican populations. Occasionally the pale green pattern along the midrib is absent but this does not appear to follow a geographic pattern, since both Chiapan and Panamanian plants show the pattern while plants from near San Salvador (El

Salvador) and Río Virilla, Costa Rica, do not. Two purple bands on the undersurface of the leaf are found in the southern populations in Nicaragua, Costa Rica, and Panamá, though plants from Monteverde and Río Virilla, Costa Rica, lacked them. The sheath is generally glabrous along the back portion in northern populations, while those in Guanacaste and near Monteverde, Costa Rica, are puberulent along the back as well.

Relationship: *Calathea macrosepala* is closely related to *C. latifolia*, sharing a similar inflorescence morphology. Both have more or less puberulent imbricate bracts, a pleated leaf, and similar capsule structure, in which abscission of the capsule is concomitant with dehiscence. The bract texture in *C. latifolia* is slightly more coriaceous. The thin herbaceous pleated leaf is also shared with *C. soconuscum*; however, the two differ with regard to the capsule morphology. The habit and the shape and morphology of the leaf are strikingly similar to those of *C. allouia*, a species found in Trinidad and the Guianas. The two species are probably distantly related.

Economic value: Standley and Calderón (1925, p. 56) report that in El Salvador "las inflorescencias [of *Calathea macrosepala*] son cocinadas y comidas y son vendidas en los mercados como verdura." In Guatemala the inflorescences are also cooked and eaten, and the leaves are used to wrap tamales (Standley and Steyermark, 1952). The inflorescences are also eaten in México (Matuda, 1950). Matuda (1950, p. 327), discussing *C. macrosepala* (as *C. allouia*), states that "después de la floración se forman en la raiz unas bolitas en forma de papas, conocidas por los nativos con el nombre de 'papitas', que también se cocinan y se comen." Standley and Calderón (1925) attribute medicinal properties to these root-tubers, stating (p. 56), "El cocimiento de las raices suprime la eliminación de azucar por la orina en los diabeticos."

Notes: Plants of *Calathea macrosepala* die back to the rhizome during the dry season, and the seeds produced at the end of the rainy season do not germinate until the beginning of the next rainy season. The structure and formation of the root-tubers in this species are discussed by Weber (1957). The shoots of cultivated plants can be prevented from dying back by constant watering during the dry season. However, in time these decline and eventually the entire plant dies. The dying of the shoots is probably triggered by soil drought, but the tubers are already present near the end of the rainy season. Their formation is not tied to water stress but may be under photoperiodic control.

Calathea macrosepala is incorrectly cited in Flora of Chiapas (Matuda, 1950), Flora of Guatemala (Standley and Steyermark, 1952) and Flora of Panama (Woodson and Schery, 1945), as *Calathea allouia*, a quite distinct plant not occurring naturally in Central America (Kennedy, 1973).

2b. *Calathea macrosepala* var. **reflexa** H. Kennedy, var. nov.

(Plate 6, Figure 7)

Quoad habitum cum *C. macrosepala* var. *macrosepala* optime congruens, sed differt interdum bracteis et sepalis purpurascentibus et folio summo angustiore et bracteis margine reflexis vel revolutis, margine adaxiali bractearum dense tomentosa, quae in *C. macrosepala* var. *macrosepala* glabra est.

Caulescent herb, 0.6–1.6 m. high. Mature shoots bearing 4–6 leaves, 3–5 basal, 1 or infrequently 2 cauline. The plant dies back to the rhizome during the dry season, sprouting again in the rainy season. Roots bearing prominent root-tubers up to 5.5 cm. long and 2.7 cm. wide. *Cataphylls* herbaceous, narrowly ovate, apex acute, mucronulate, green, subglabrous. *Leaf blade* thin, herbaceous (dried, membranaceous), the major lateral veins raised, giving a pleated appearance; upper leaves narrowly elliptical, acute to obtuse; lowermost leaf or couple of leaves ovate to elliptic, apex obtuse with an acumen to acuminate; base obtuse; 41–76 cm. long and 11.4–18.9 cm. wide in subtending leaf; leaves 2–4, 32.5–68.2 cm. long and 9.5–18.9 cm. wide; lowermost 16–19 cm. long and 7.3–8.9 cm. wide. The subtending leaf or the penultimate leaf is the largest, the other leaves decreasing in size down the shoot. Leaf above slightly glittering, grass green, glabrous, midrib area bearing a narrow, feathery, pale yellow green pattern, this more pronounced in younger plants and lower leaves on a shoot, commonly 1.5 cm. to occasionally 2 cm. wide, midrib yellow green, puberulent; lower surface opaque pale grey green, veins appearing darker green, bearing two longitudinal bands of light purple, especially noticeable in young plants, paler in upper leaves on a shoot, glabrous, midrib yellow green, glabrous to subglabrous. *Pulvinus* elliptical in cross section, yellow green, glabrous occasionally with minute hairs along the upper side, junction with petiole marked by a pale yellow green ring, the pulvinus acuminately extended up to 1.5 cm. along the upper surface of the petiole, (0.8)1.3–4 cm. long in subtending leaf, in second to third leaves 1.3–3.4 cm. long, in lowermost leaf, 0.5–0.6 cm. long. *Petiole* deep green, glabrous, bearing a groove on the upper side *ca.* 1.5 mm. wide, 3.5–28.5 cm. long in subtending leaf; others, 0–18 cm. long. *Leaf sheath* not auriculate in upper leaves, sheath of the first (occasionally first and second) leaf on a shoot, auriculate; herbaceous, glabrous except for puberulent apical margin. *Stem* green, glabrous.

Inflorescence terminal, subtended by a foliage leaf, ovoid to narrowly ovoid, imbricate, 4.5–7.2 cm. high and 3–4.6 cm. wide. *Peduncle* green or tinged purple just below junction with bracts, glabrous, (4.5)7–25.5 cm. long. *Bracts* spiral, 12–30, semicoriaceous, depressed ovate-trullate, obtuse to 90°, infrequently lowest bract acuminate, base imbricate, margin reflexed to revolute, 1.9–2.8 cm. high and 2.9–4.1 cm. wide, subtending up to 7 flower pairs. Outer surface light green, less commonly purple or green with upper half tinged purple, minutely puberulent; inner surface shiny pale cream green, glabrous except for puberulent recurved apical margin. *Bicarinate prophyll* membranaceous, transverse broadly elliptic, apex rounded, faint translucent green above, semitransparent below or tinged purple at apex, subglabrous to puberulent apically, 1.5–1.9 cm. high and 1–1.55 cm. wide, carina to carina, 1.6–2.2 cm. total width. *Secondary bracts* membranaceous, transverse broadly obovate-elliptic, more or less transparent or tinged purple at apex, glabrous to puberulent at apex, 1.5–1.8 cm. high and (1.5)1.7–2.3 cm. wide. *Bracteoles* 2 per flower pair, membranaceous, channeled or carinate, transparent, glabrous, 1.4–1.7 cm. high and 0.3–0.5 cm. wide.

Flower never opening spontaneously. *Sepals* channeled, herbaceous, lateral margins hyaline, narrowly elliptic to narrowly obovate, acute to obtuse, in bud the 3 sepals meeting to form an acute point, white basally, faint green to greenish yellow or tinged purplish above, glabrous, 2–2.45 cm. high and (0.45) 0.5–0.75 cm. wide. *Corolla tube* white at base, cream to pale yellowish above, glabrous, (2.7)2.8–3 cm. long; lobes unequal, elliptic to broadly elliptic, obtuse, appearing acute due to inrolled margins, yellow at base, tips light yellow green, 1.4–1.75 cm. long and 0.55–0.85 cm. wide. *Outer staminode* cuneiform, apex shallowly dished, rounded to emarginate, faint orangish, 1.1–1.2 cm. long and 0.3–0.65 cm. wide. *Callose staminode* rectangular, apex bi- or tridentate to one side, light orange (apricot) 0.95–1.1 cm. long. *Cucullate staminode* cream-colored apically, basal portion golden orange, 1–1.1 cm. long and 0.4–0.45 cm. wide, bearing a fingerlike channeled appendage subterminally to 1.5 mm. high. *Filament* with a lateral petaloid appendage to 1.5 mm. wide; anther 2–2.5 mm. long. Style and stigma golden. *Ovary* white, glabrous 1.5–2.3 mm. high. *Capsule* obovoid, rounded apically, straw-colored, 0.9–1 cm. high and 0.55–0.65 cm. in diameter, abscising from the pedicel at the time of dehiscence, the tissue at the base is swollen and mealy in appearance at the time of abscission. The capsule is crowned by a persistent enlarged calyx to 0.8 cm. wide. *Seeds* usually 3 per capsule, trigonous, rugose on outer surface, slate grey, *ca.* 0.45–0.5 mm. high and 3.5–4 mm. in diameter, bearing a basal white aril 2.5–3 mm. high.

Type: COSTA RICA. Prov. Puntarenas: on west facing slope of mountain overlooking Villa Neilly, along roadside on limestone outcrop, *ca.* 100 m., flowered in cultivation, 24 Oct. 1970, *H. Kennedy 650* (holotype F!, isotypes CR!, US!).

Specimens seen: COSTA RICA. Puntarenas: on west facing slope of mountain overlooking Villa Neilly, along roadside on limestone outcrop, *ca.* 100 m., 25 June 1969, *H. Kennedy 267* (MO); *ibidem* flowered in cultivation 30 June 1969, *H. Kennedy 26,* (CR, MO); entre Villa Neilly y Agua Buena, Golfito, 500 m., 21 Oct. 1964, *A. Jiménez M. 2449* (F, MO). San José: along the Pan Am Highway, about halfway between San Isidro del General and Cerro de la Muerte, *ca.* 1400 m., 26 Aug. 1968, *H. Kennedy and R. B. Foster 72* (US); *H. Kennedy and R. B. Foster 73* (CR).

Flowering: *Calathea macrosepala* var. *reflexa* flowers during mid to late rainy season, June through October, dying back to the rhizome during the dry season.

Pollinator: Plants of *Calathea macrosepala* var. *reflexa* grown in cultivation near San Vito, Costa Rica, were observed to be pollinated primarily by *Eulaema cingulata* (Fabricius) and infrequently by *E. speciosa* (Mocsary), while populations near Villa Neilly were observed to be pollinated only by *E. cingulata* (one day's observation).

Distribution and habitat: *Calathea macrosepala* var. *reflexa* occurs mainly on the Pacific slopes of the Talamanca Range in Costa Rica, primarily on limestone outcroppings. It is found from Villa Neilly, Puntarenas Province in the south, northward to above San Isidro del General along the Pan American Highway. It is commonly found in open exposed areas on moderate to steep slopes, a prevalent colonizer of the steep road cut along the road between Villa Neilly and Agua Buena, not occurring within the forest.

Variation: *Calathea macrosepala* var. *reflexa* is characterized by the recurved to revolute margins of the bracts. The recurved margin is densely puberulent (seen under magnification), the hairs being longer than those on the lower portion of the bract. The bracts are minutely puberulent, the hairs being shorter and less dense than in the var. *macrosepala*. The subtending leaf tends to be proportionally narrower than those of var. *macrosepala*, the ratio of width to length is 1:3.5-1:4 in var. *reflexa* and 1:1.9-1:3.6 in var. *macrosepala*. The base of the leaf sheath is minutely puberulent to subglabrous, whereas in var. *macrosepala* it is densely puberulent to appressed tomentose. The pubescence on the lower surface of the leaf is less dense and of shorter hairs than var. *macrosepala*. In the population near Agua Buena and above San Isidro, individuals with purple bracts and sepals occur in addition to those with green bracts and cream sepals. The ovary, capsule and seeds are smaller in var. *reflexa* than var. *macrosepala*.

One collection (*Skutch 2688*) in the vicinity of El General in San José Province was found to be nearly intermediate between the two varieties.

Notes: The seeds are produced during the latter part of the rainy season, remaining dormant through the dry season, and germinate at the beginning of the following rainy season.

3. *Calathea latifolia* (Willdenow ex Link) Klotzsch

(Plate 7, Figures 8, 9, 10)

Alpinia latifolia Willdenow ex Link, Jahresber. 1:22-23. 1820. Type: ad Caripe in umbrosis, Venezuela, *Humboldt and Bonpland 203* (B—WILLDENOW Herb. 20). Locality *fide* Koernicke (1862), collection no. *fide* Schumann (1902). *Thalia latifolia* Link ex Schultes, Mantissa 1:10. 1822. *Calathea latifolia* (Willd. ex Link) Klotzsch, *in* R. Schomburgk, Reisen Brit. Guiana 3: 918. 1848. *Phyllodes platyphyllum* O. Kuntze, Rev. Gen. Pl. 2: 696. 1891. Based on *Thalia latifolia* Link ex Schultes not *Phrynium latifolium* Blume.

Calathea allouia var. *violacea* (Lindley) Woodson, Ann. Missouri Bot. Gard. 29(4): 332. 1942. *Sensu* Woodson, *non Calathea violacea* Lindley. Cited by Woodson as var. *violacea* (Roscoe) Woodson.

Cauline herb (0.68)0.87-2.26 m. high. Shoots bearing 3-7 leaves, (1)2-5 basal, 1-4 cauline. Roots bearing root-tubers, the plant dying back to the rhizome during the dry season. Cataphylls 1-3, firm at the base to herbaceous at apex, ovate to narrowly ovate, obtuse to acute, mucronulate, green, striped with purple to wholly purple, scattered puberulent, the innermost one, 16-48 cm. long. *Leaf blade* herbaceous, ovate to narrowly ovate or oblong-ovate, apex obtuse with a point, base truncate to rounded, the connection of the blade to pulvinus shortly acuminate, subtending leaf 14-57 cm. long and 8.7-24.5 cm. wide; others,

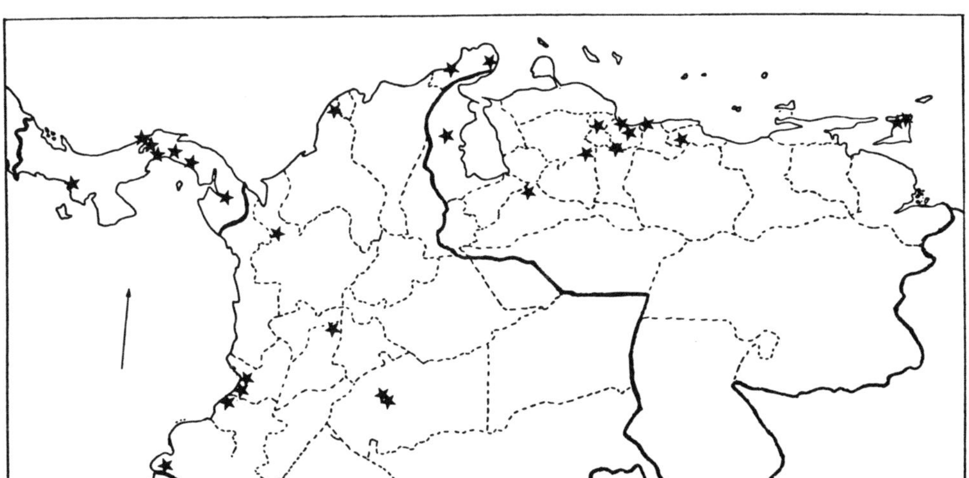

Fig. 8. Distribution of *Calathea latifolia* in Panamá, Colombia, Venezuela, and Trinidad.

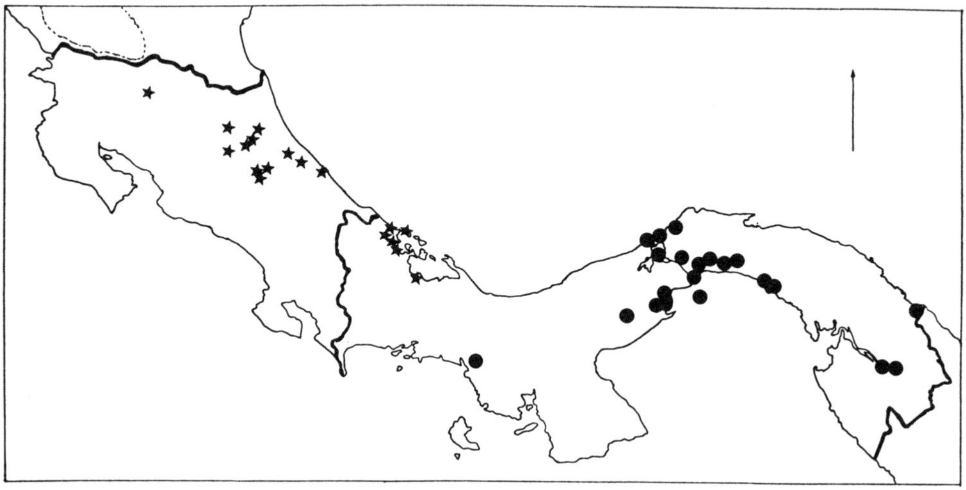

Fig. 9. Distribution of *Calathea warscewiczii* (stars) and *C. latifolia* (circles) in Costa Rica and Panamá.

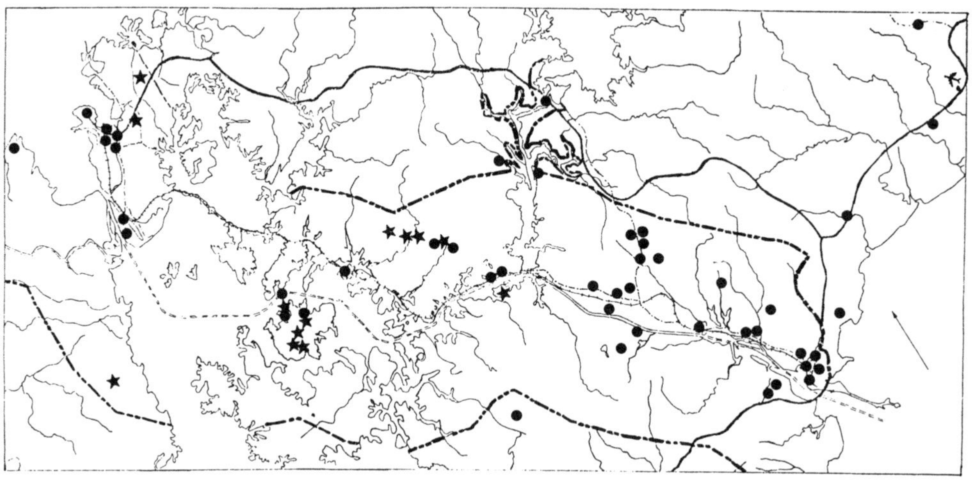

Fig. 10. Distribution of *Calathea latifolia* (circles) and *C. marantifolia* (stars) in the Panama Canal Zone.

24–74 cm. long and 10.8–33 cm. wide. Leaf surface pleated, major lateral veins raised, glittering, bright green, glabrous, except for the midrib which is subglabrous basally to puberulent toward apex; lower surface opaque pale greyish green, frequently bearing two longitudinal broad purple bands along the blade between the midrib and the margins, glabrous to minutely puberulent, midrib yellow green. *Pulvinus* yellow green, often streaked with purple, glabrous to minutely puberulent throughout, pubescence more pronounced along upper surface and at junction with blade, (0.7)2–7.5 cm. long, junction with petiole articulate, extension of pulvinus acuminate above and below, marked by a pale greenish yellow band. *Petiole* grass green throughout or streaked with purple, subglabrous to densely minutely tomentose, 4–32 cm. long in subtending leaf, occasionally absent; others (0)7.8–71 cm. long. *Leaf sheath* coriaceous to chartaceous, apex not auriculate, green or green streaked with purple, purple striping increasing from apex to base of sheath, the nodal area frequently entirely purple and also commonly the lower leaf sheaths, subglabrous to densely minute puberulent medially above, margins and lower portions subglabrous or puberulent to densely velvety tomentose; within, pale green to greenish white, glabrous, 10-37.5 cm. long in subtending leaves; others 22-109 cm. long. *Stem* grass green or green streaked with purple, minutely tomentose to glabrous.

Inflorescence terminal, subtended by a foliage leaf, imbricate, ovoid to cylindrical, 5.5-14 cm. high and 3.5-5 cm. in diameter. *Peduncle* green, subglabrous to minutely puberulent, especially at junction to the inflorescence, 8-37(47) cm. long. *Bracts* (9)13-53, spiral, coriaceous, apex slightly out-turned, reniform to depressed ovate-elliptical, apex rounded to broadly obtuse, 1.8-2.6 cm. high and 2.6-4.65 cm. broad. Outer surface wholly green, basal 1/2-2/3 green with apical portion streaked or flushed with purple or entirely purple, densely minute puberulent; within, pale green, shiny, glabrous, lateral margins hyaline. Each bract may subtend up to 13 flower pairs. *Bicarinate prophyll* membranaceous, rotund, apex obtuse to rounded, translucent green, or green tinged purple at apex, glabrous or minutely puberulent at apex only, 1.7-2.4 cm. high and 1.6-2.6 cm. wide. *Secondary bracts* membranaceous, broadly obovate-oblong, apex truncate to subretuse, translucent pale green or tinged purple at apex, glabrous, 1.5-2.25 cm. high and 1.6-2.3 cm. wide. *Bracteoles* subtending individual flowers membranaceous, translucent white below, pale green above, glabrous, 0.9-1.9 cm. high.

Flowers never open spontaneously. *Sepals* membranaceous at the margins, semispongy basally and medially, channeled, narrowly oblong-obovate, apex acute, shiny, pale green or with lower half white and purple above, glabrous, 1.65-2.45 cm. long and 0.6-1.1 cm. wide. *Corolla tube* white below, tinged cream or purple above, 1.9-2.9 cm. long; corolla lobes unequal, obovate-elliptic, apex obtuse, shiny, dark purple, lavender, cream or pale yellowish, 1.2-1.9 cm. long and 0.6-0.9 cm. wide. *Outer staminode* obovate, dished at apex, apex irregularly rounded or bilobed, white, occasionally with a streak of purple, 0.9-1.2 cm. long and 0.45-0.75 cm. wide. *Callose staminode* rectangular, apex truncate, irregular, margin often purple, 0.85-1.1 cm. high. *Cucullate staminode* cream to cream white, basal portion golden, 0.9-1 cm. long and 0.45 cm. wide, appendage on hood *ca.* 1.5 mm. long. *Filament* with reduced lateral petaloid appendage to 1.5 mm. wide; anther 3 mm. long. Style and stigma golden. *Ovary* white, glabrous, 2-3.5 mm. high and 2-2.5 mm. in diameter. *Capsule* thin, smooth, obovoid, rounded at the apex, pale cream color, apical portion frequently tinged purple, 1-1.3 cm. high and 0.7-0.9 cm. in diameter, the tissue at the very base is mealy at time of abscission of the capsule. The capsule is crowned by a live, persistent, dark purple, green, or green apically tinged purple, calyx; sepals enlarged to 1.2 cm. wide. *Seeds* usually 3 per capsule, trigonous, rugose on outer surface, slate grey 5-6 mm. long and 4-4.8 mm. in diameter; bearing a basal white aril, 2.5-4 mm. high.

Representative specimens seen: TRINIDAD: Valley road, Belmont, under trees, 27 Aug. 1908, *W. E. Broadway 2359* (BR); Chancellor Road, bottom part in low lying wet ground, 4 Sept. 1929, *W. E. Broadway 7399* (BM, F, MO, S, US); Laventille, near the main road, in partial shade, 7 Sept. 1931, *W. E. Broadway 7737* (BM); St. Ann's, Sept. 1845, *Crueger s.n.* (TRIN); Maracas Valley Road, mile post 1.5, 29 Sept. 1972, *H. Kennedy, Bhorai, Ramkissoon 1700* (BM, DAV, DUKE, F, MO, TRIN, U, UC, US, VEN); St. Augustine ICTA orchard among grass, alt. low, 30 July 1945, *N. W. Simmonds 11* (K, TRIN); Maracas Valley near Marsden's Pool, *ca.* 200 m., 12 Aug. 1945, *N. W. Simmonds 12* (TRIN); St. Augustine, ICTA, orchard among grassy citrus, alt. low, 5 Sept. 1947, *N. W. Simmonds 167* (TRIN); without locality, 1877-1880, *A. Fendler 797* (BM, K, NY).

VENEZUELA. Monagas: without exact locality, 9 Sept. 1950, *F. D. Smith 199* (US). Miranda: Ocumare del Tuy, orillas de la quebrada Mararito, 270 m., 14 Sept. 1964, *F. Matos 1053* (VEN). Aragua: Ocumare de la Costa, in cacao plantations, 31 July 1925, *Pittier 11850* (NY, US). Carabobo: alrededores de Valencia en lugares sombrios, 15 Aug. 1920, *H. Pittier 9047* (GH, NY, US, VEN), label read Aragua for GH, US, and VEN, but NY reads Carabobo. Yaracuy: Hacienda Iboa, cerca de Guama, 4-500 m., 13 Sept. 1923, *H. Pittier 11151* (GH, US, VEN). Cojedes: alrededores de Tinaquillo, Hacienda Aquirre, sitios sombrios, July 1960, *L. Aristequieta 4307* (COL, VEN). Portuguesa: selva torpófila sobre peñasco calcáreo, 5 km. E.N.E. de Agua Blanca, 22 km. N.E. de Acarigua, 190 m., 24 Aug. 1966, *J. A. Steyermark and M. Rabe 96450* (F, US, VEN). Barinas: near Barinitas, secondary regrowth, 500 m., 2 Sept. 1964, *F. J. Breteler 4174* (NY, U, US). Lara: Montañita de Cárcamo, 3 Aug. 1930, *J. Saer 608* (VEN). Zulia: Serra de Perija, faldas infer., a lo largo del Río Yasa, vecindad de "Guasama," arriba de "Kasmera" (Est. Biol. de la Univ. del Zulia), al S.O. de Machiques, 500-600 m., 26-27 Aug. 1967, *J. A. Steyermark and J. E. Fernández 99819* (VEN). Guarico: Ortíz-Galeras de El Pao, a unos 20 km. de San Francisco de Tiznados, Hato El Tranquero del Dr. H. Garrido, 20 Aug. 1966, *L. Aristequita, G. Agostini, E. Castellano 6280* (VEN). State unknown: La Boca, S. of Santa Lucia, 5 Aug. 1891, *Eggers 13548* (S, US).

COLOMBIA. Guajira: Serrania de Macuira, veriente S.O., cercanias de la quebrada de Taquaira, 400 m., 3 Dec. 1953, *R. R. Casañeda 4486* (COL). Magdalena: near Manaure, 700 m., 1 Sept. 1944, *O. Haught 4338* (BM, COL, F, US); foothill of the Serra de Perijá, Manaure, 750-790 m., marginal forest of Río Manaure, 4 Nov. 1959, *J. Cuatrecasas and R. Romero Castañeda 24996* (COL, US); 8 km. E. of Codazzi, on Quebrada Mansa, 2 Sept. 1938, *O. Haught 2312* (COL, GH, US); Santa Marta, near "La Miranda," *ca.* 10 m., 8 Aug. 1926, *A Schultze 512* (B); Santa Marta, 1000 ft., Nov. 1898-1901, *H. H. Smith 2425* (BM, F, GH, K, MICH, NY, P, US). Bolívar: vicinity of Turbaco, 200-300 m., 6-22 Nov. 1926, *E. P. Killip and A. C. Smith 14293* (NY, US). Antioquia: Urabá, Municipio de Mutatá, Villa Arteaga, *ca.* 150 ft., 16-20 Feb. 1953, *R. E. Schultes and I. Cabrera 18645* (COL); selva pluvial y abierta, carretera del mar en los alrededores del Río Ampurrumiadó, ppt. pluvial 3750-5000 mm. anuales, *ca.* 200 m., 11 Oct. 1947, *G. Gutiérrez V. and R. A. Barkley 17C215* (COL, GH). El Valle: costa del Pacífico, Río Cambre, Barco, 5-80 m., 21-30 Apr. 1944, *J. Cuatrecasas 17233* (F, US); secondaria, cerca campamiento de RF virus estación Río Raposo, 20-30 m., 5 May 1966, *V. H. Lee 64* (COL). Cauca: costa Pacífico, Río Micay, brazo Noanamito, orilla derecha, El Chachajo, 2-5 m., 27 Feb. 1943, *J. Cuatrecasas 14243* (F, US). Nariño: 4 leguas al oeste de La Espriella, 8 July 1952, *R. R. Castañeda 3164* (COL); Isla de Gallos Pacífico, campos de este, Feb. 1942, *Dryander 2584* (US); costa de Pacífica, corregimiento Herrera, selva pluvial en la cuenca del Río Mira, cerca del cacerio de Candelillas, 30-100 m., 27-29 April 1953, *J. M. Idrobo and H. Weber 1370* (COL). Meta: Río Guëjar *ca.* 10 km. below junction with Río Zanza, N. end of Cordillera Macarena, dense forest, in deep shade, 470 m., *S. Galen Smith 1517* (COL, GH, US); dense forest in deep shade, Río Guëjar *ca.* 10 km. below junction with Río Zanza, N. end of Cordillera Macarena, 470 m., 21 Aug. 1950, *S. G. Smith and J. M. Idrobo 1517* (COL); Río Guëjar, forest S. of El Mico Airstrip (Camp 1) 400 m., 6 Nov. 1949, *W. R. Philipson, J. M. Idrobo, A. Fernández 1345* (BM, COL, S, US). Without exact locality: Nelle Grenade, 1844, *M. J. Goudot 134* (P); Amer. Aequinoctialis (Magdalena) *E. André 462* (BM, K).

PANAMÁ. Darién: vicinity of Yape, *ca.* 30 m., 4 Oct. 1938, *P. H. Allen 861* (GH, MO). San Blas: vicinity of Pto. Olbaldía, 19 Aug. 1971, *T. Croat 16886* (MO). Panamá: vicinity of Río Tapia, east of Panama City, near Panama National Highway, 2 Aug. 1940, *H. H. Bartlett and T. Lasser 16619* (MICH, MO); along Pan Am Highway, open grazed area near Jenine, Río Cañita, 24 Sept. 1961, *J. A. Duke 3829* (MO); Pacora, 30 July 1961, *J. D. Dwyer 1256* (NY); Icanti, Río Agua Clara, 19 Nov. 1971, *A. Gentry 2612* (MO); Perlas Archipelago, San José Island *ca.* 55 mi. S.S.E. of Balboa, 18 Oct. 1944, *I. M. Johnston 208* (GH, US); *ca.* 2 mi. up the road to Cerro Jefe from the Pan Am Highway, 26 July 1970, *H. Kennedy and R. L. Dressler 425* (DAV); Chepo, 1 mi. from the river along roadside, 12 Aug. 1970, *H. Kennedy 440* (K, DAV); along Pan Am Highway *ca.* 1 mi. S.W. of the Capira guard station, in a field, 23 Oct. 1971, *H. Kennedy 1165A* (F); along road up to Cerro Campana semideciduous forest, secondary growth along roadside, 400-500 m., 20 Oct. 1971, *H. Kennedy 1172* (US); Río Pasiga, beach area near the mouth and secondary vegetation *ca.* 1 km. back from the beach, 0-10 m., 26 Oct. 1971. *H. Kennedy 1178* (MO); Río Pasiga, *ca.* 3-4 mi. upstream, semideciduous forest, 28 Oct. 1971, *H. Kennedy 1199* (MO); *H. Kennedy 1221* (DAV, UC); Bella Vista, at sea level, 16 Oct. 1922, *E. P. Killip 12032* (GH, US); S.E. slope of Cerro Campana, roadside and grassland, 10 Dec. 1967, *W. H. Lewis, W. H. Blackwell Jr. et al. 3141* (MO); off Road k-15 near Huile on the hills around Gatun Lake, secondary and cut-over primary forest, 26 Dec. 1957, *C. E. Smith Jr., H. M. Smith, T. Arruaz 3288* (F, US); near big swamp E. of the Río Tocumen, wet forest, 11 Dec. 1923, *P. C. Standley 26676* (US); along Corazal Road near Panamá, 13 Dec. 1923, *P. C. Standley 26839* (US); Taboga Isl., Dec. 1923, *P. C. Standley 27942* (US); Chepo, near river, 3 Oct. 1965, *E. Tyson 1416* (SCZ); marsh area 2 mi. S. of Tocumen Airport, 25 July 1969, *Tyson and Clewell 5916* (DUKE, SCZ); *E. L. Tyson and A. F. Clewell 5917* (DUKE); Agri. Exp. Station at Matías Hernandez, 20 Aug. 1914, *H. Pittier 6753* (GH, US); *H. Pittier 6754* (US); Camino de Pacora, 10 Aug. 1969, *C. Wendehake 21* (MO); Río Tatare, 3-20 Aug. 1940, *R. E. Woodson, Jr. and R. W. Schery 1010* (MO, NY); *R. E. Woodson, Jr. and R. W. Schery 1011* (BM); boggy grasslands and marginal thicks, between Pacora and Chepo, *ca.* 25 mi., 1 Aug. 1938, *R. E. Woodson Jr., P. H. Allen, R. J. Seibert 1669* (GH, MICH, MO, NY); *R. E. Woodson, Jr., P. H. Allen, R. J. Seibert 1670* (GH, MICH, NY); Cerro Campana, near Su Lin Motor Lodge, *ca.* 3000 ft., 31 July 1967, *F. Woods, A. Woods, P. Woods s.n.* (DUKE). Canal Zone: Juan Mina, 22 Aug. 1940, *H. H. Bartlett and T. Lasser 16811* (MICH, MO);

Orchid Garden Ancon Hill, 31 July 1940, *H. H. Bartlett and T. Lasser 16611* (DUKE, MICH, MO); Fort Clayton, out C-15 road, 15 Aug. 1966, *K. E. Blum and J. Dwyer 2686* (MO); Barro Colorado Isl., ruins of French Canal, 19 Aug. 1928, *A. M. Chickering 24* (MICH); clearing at Rear No. 8 lighthouse, 23 Sept. 1969, *T. Croat 6351* (MO, SCZ); rear lighthouse clearing, 7 Sept. 1970, *T. Croat 11989* (BCI HERB., MO); Mohinga swamp, 15 July 1931, *L. H. Bailey and E. Z. Bailey 580* (F); Ft. San Lorenzo, 17 July 1960, *J. E. Ebinger 475* (US); Madden Dam, 10 Aug. 1960, *J. E. Ebinger 838* (MO); Navy Reservation (Pipeline Road) N. of Gamboa at marker #75, 21 Aug. 1971, *H. Kennedy 454* (F); Ancon Hill beside bld. #235 (STRI), 22 Oct. 1971, *H. Kennedy 1173* (COL, DAV, F, GH, K, NY, PMA, U, US); N. of Frijoles, in dense forest, 20 May 1918, *E. P. Killip 3430* (US); Ancon Hill, 100-200 m., 17 Oct. 1922, *E. P. Killip 12081* (NY, US); around Culebra, 50-150 m., 1911, *H. Pittier 4061* (US); Ancon Hill, 8 Sept. 1920, *W. W. Rowlee and H. E. Stork 976* (US); Foresta Madden, 23 Sept. 1960, *D. Sucre B. 81* (RB); Ancon Hill, 26 Nov. to 9 Dec. 1923, *P. C. Standley 25164* (US); Sousa Hill, Balboa, 27 Nov. to 10 Jan. 1923, *P. C. Standley 25224* (US) and *26477* (US); Balboa, moist thicket, Nov. 1923 to Jan. 1924, *P. C. Standley 25441* (US); Las Cascadas Plantation, near Summit, 2 Dec. 1923, *P. C. Standley 25709* (US); Cerro Gordo near Culebra, 5 Dec. 1923, *P. C. Standley 25968* (US); along the old Las Cruces trail, between Fort Clayton and Corozal, 31 Dec. 1923, *P. C. Standley 29065* (US); Frijoles, 20 Aug. 1923, *F. L. Stevens 1197* (US); Chiva Chiva trail near Miraflores Lake, 30 Sept. 1965, *E. L. Tyson 1359* (MO, SCZ). Colón: Colón, 6 Oct. 1960, *D. Sucre B. 244* (RB); vicinity of Palenque, near sea level, 7-8 Aug. 1911, *H. Pittier 4129* (US); María Chiquita, E. of Río Piedras toward Portobelo, thickets near beach, 9 Aug. 1967, *J. D. Dwyer and J. H. Kirkbride Jr. 7785* (GH, MO, US). Coclé: El Valle de Antón in cut-over field, 3 Aug. 1970, *J. L. Luteyn and H. Kennedy 1715* (DUKE). Chiriquí: between Río Tinta y Río Tabasara along main highway, 8 Nov. 1937, *Woodson, Siebert, Allen 425* (MO). Without exact locality: 4 Aug. 1876, *E. André K590* (K, NY); 1859-1860, *S. Hayes 227* (NY); Sabanas, July 1932, *Paul 103* (US).

Flowering: *Calathea latifolia* flowers in the rainy season, from late July through November in Panamá, Atlantic coastal Colombia, Venezuela, and Trinidad. It has a similar flowering time in the Intendencia of Meta, Colombia; however, plants from El Valle, Cauca, and Nariño, Colombia were collected in flower in February, April, and May.

Pollinators: *Calathea latifolia* is pollinated primarily by *Eulaema cingulata* (Fabricius) and secondarily by *Euglossa imperialis* Cockerell on Pipeline Road, Canal Zone, while *Euglossa imperialis* is almost the exclusive pollinator on Barro Colorado Island, Canal Zone. In the drier habitat on Ancon Hill, Canal Zone, it is pollinated by *Eulaema nigrita* Lepeletier. Populations along the Maracas Valley road, Trinidad, were pollinated by *Eulaema cingulata.*

Distribution and habitat: *Calathea latifolia* occurs from western Panamá south along the drier Pacific side into the Darién and on the Atlantic side near Puerto Obaldía and is found in the seasonal semi-deciduous forest zone along the Atlantic coast of Colombia and Venezuela to Trinidad. To the south it occurs along the Pacific lowlands of Colombia in El Valle, Cauca, and Nariño, and also occurs inland in the Meta Valley near the Cordillera de Macarena. I have not seen any specimens from Guyana though Klotzsch (1848, p. 918) reports this species from there when transferring it to the genus *Calathea.* This species occurs in seasonal deciduous to semideciduous forest and in Colombia is often in a gallery forest. The plants usually grow at the forest edge and in the adjacent open, very often disturbed, areas. In the drier areas they may occur within the forest, but the forest is quite open and the shading minimal. This species is quite intolerant of shade and is common in more open areas. On the Atlantic side of the Canal Zone, it forms dense stands on exposed cliff edges overlooking the beach, almost 10 meters from the sea.

The collections in El Valle, Cauca, and Nariño are somewhat anomalous and may represent a variant form as the Nariño plant is cited as growing in "selva pluvial" (*Idrobo and Weber 1370*), and the flowering times suggest a wetter habitat.

Variation: The general habit of *Calathea latifolia* is uniform throughout the range, however the average height of the plant and size of the leaves, especially the subtending leaf, decrease clinally from Panamá to Trinidad; the plants from the Meta Valley, Colombia, are slightly smaller also. The pubescence on the underside of the leaf is variable even between nearby populations. However, the majority of the plants have a glabrous to subglabrous underleaf while nearly all individuals in Trinidad have a puberulent underleaf. The leaf sheath is densely puberulent to tomentose or subvillose throughout in Panamanian populations, minutely puberulent in the Atlantic Colombian, Venezuelan, and Trinidad populations. The apex and margins are minutely puberulent and the rest subglabrous to glabrous in the Meta Valley plants. The bracts of Panamanian plants are more densely pubescent than are those of Trinidad plants, while populations from the Meta Valley, Colombia have glabrous bracts. Plants from Ancon Hill, Canal Zone, have a glabrous pulvinus whereas those in Trinidad have a puberulent pulvinus. The Trinidad plants have generally fewer bracts and the inflorescence is not so elongate (5.5–8 cm. high) as in the Panamanian individuals (6–14 cm. high). The color variation has been previously noted by Woodson (Woodson and Schery, 1942) for Panamanian plants, which led to occasional confusion of *Calathea marantifolia* with the cream colored flower form (Kennedy, 1973). Within Panamá the two main color variants are the purple-flowered form and the cream-flowered form. In Panamá, the cream-flowered individuals also lack purple markings on the leaves and stems and have wholly green bracts. Occasionally the petals are lavender, being much paler in tint than the usual purple. More rarely, mottled petals are encountered in which the petals are dotted with areas of purple and cream. Mixed populations are common near Chepo and around Chimán, Panamá, the cream form being very rare in the Canal Zone. Plants from Puerto Obaldía, Panamá, eastward to Trinidad have cream to pale yellowish petals and purple or purple tinged bracts.

Relationship: *Calathea latifolia* is closely related to *Calathea macrosepala* and to *C. marantifolia. Calathea macrosepala* and *C. latifolia* have a thin herbaceous, pleated leaf, the capsule abscises at the time of dehiscence, and its base is swollen and mealy textured. Both species die back to the rhizome at the beginning of the dry season and have large root-tubers. The habit, bract shape, inflorescence morphology, and flower morphology are similar to those of *C. marantifolia.* The sepals are more coriaceous in *C. latifolia* and are rounded at the apex in bud as are those in *C. marantifolia,* as opposed to the semi-herbaceous, acutely pointed sepals of *C. macrosepala.* Schumann placed *C. latifolia* in the series *Nudiscapae* Petersen, but it properly belongs in *Scapifoliae* Eichl. *apud* Petersen.

Economic value: The closely related *Calathea allouia* (lirén) is cultivated in the West Indies for the edible root-tubers, and early reports, according to Sturtevant (1969), mention the Indians using a "lirén-like" plant also for food. "The other plant . . . is one which Peter Martyr called 'cabaióes' . . . ; Oviedo wrote the same Taino name as 'cauallos,' which he says resembles the lirén but are somewhat larger, 'grow in poor and thin soils . . . [and are] wholesome food and enjoyed by the Indians. It is a wild fruit and it comes up and grows through the care of nature alone' " (Oviedo y Valdés, 1851, quoted in Sturtevant, 1969, p. 185). My observations of *C. latifolia* in Trinidad are in complete agreement with Oveido's remarks about the plant, and it is most probable that *C. latifolia* is indeed this "other" plant.

Notes: Four artificial self-pollinations of *Calathea latifolia* from Pipeline road, Canal Zone, and three of *C. latifolia* under cultivation from the Meta Valley, Colombia, were made. Six of these set seed, indicating that this species is probably self-compatible. Seeds collected at the time of capsule dehiscence from a population on Pipeline Road, were planted in a greenhouse and did not germinate until after the dry season. There was almost 100% germination shortly after the beginning of the rainy season. The seeds were not watered, except through rain, during the dry season.

In the *Flora of Panama,* Woodson and Schery (1945) misidentified *Calathea latifolia* as *C. allouia* (Aublet) Lindley, a closely related cultivated species not occurring in Panamá, and cited it as the trinomial *C. allouia* var. *violacea* (Lindley) Woodson. However, Woodson does not refer directly to Lindley's name but to *Phrynium violaceum* of Roscoe, overlooking the fact, as many authors have, that Roscoe's (1828) name is a new combination based on *Calathea violacea* Lindley (1826a).

4. *Calathea marantifolia* Standley

(Plate 8, Figures 6, 7, 10, 11)

Calathea marantifolia Standley, Jour. Washington Acad. Sci. 17(10): 250. 1927. Type: Wet forest, El Arenal, 485–600 m., Provincia de Guanacaste, Costa Rica, 18–19 Jan. 1926, *P. C. Standley and J. Valerio 45310* (holotype US 1254038!).

Calathea lagunae Woodson, Ann. Missouri Bot. Gard. 29: 333. 1942. Type: Western River, Bocas del Toro Province, Panamá, 19 Sept. 1941, *H. von Wedel 2706* (MO 1240046!; isotypes BM!, GH!, US 1823446!).

Calathea violacea var. *hirsuta* Petersen, Fl. Bras. 3(3): 114. 1890. Lectotype (here chosen): Gatun, Panamá as "Gatum in Panama," Jan. 1858, *Dr. M. Wagner s.n.* "Fasc. 1" (M!).

Rhizomatous herb, (0.57)0.7–2.31 m. high. Shoot bearing 2–7(9) leaves, 0–3(5) basal and 2–4(5) cauline. Roots bearing root-tubers, more prevalent in juvenile plants, often poorly developed in adults. *Cataphylls* usually 3, occasionally 2, chartaceous, thicker and fleshy toward base, narrowly ovate, acute, mucronulate, green or green streaked purple at base, the shorter two frequently wholly purple toward base, the lower

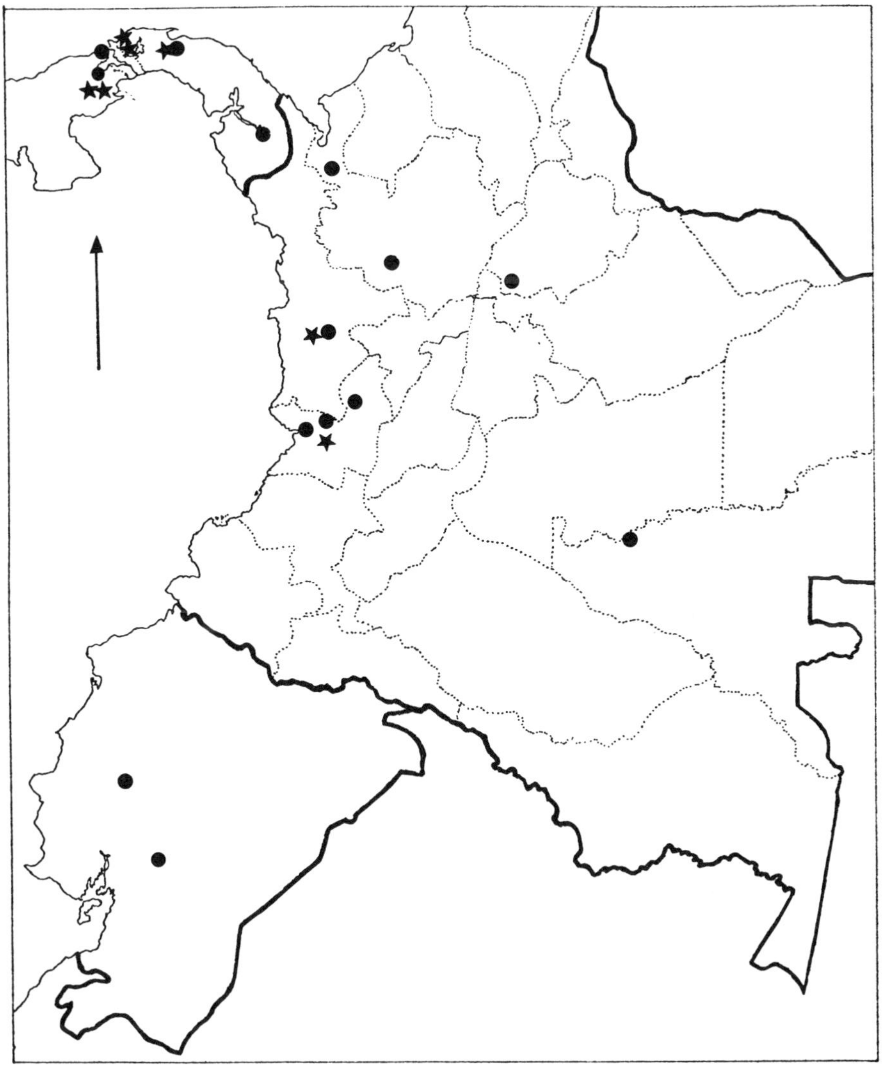

Fig. 11. Distribution of *Calathea marantifolia* (circles) and *C. silvicola* ssp. *silvicola* (stars) in Panamá, Colombia, and Ecuador.

ones and upper half of the innermost cataphyll dying, becoming thin and scarious; innermost cataphyll, (12)34.5–98 cm. long. *Leaf blade* shallowly curved between the raised major veins, ovate to narrowly ovate or less often elliptic (type, narrowly elliptic), apex obtuse to rounded with an acumen, or abruptly acuminate in smaller leaves or individuals, occasionally acuminate, base broadly obtuse or rounded to truncate, junction with the pulvinus abruptly short acuminate, 10–69 cm. long and (3.8)6.3–33 cm. wide in subtending leaf; others 14.7–74.5 cm. long and 8.2–37 cm. wide. Upper leaf surface slightly sparkling in younger leaves, most frequently opaque, slightly dullish in appearance, deep green, glabrous, midrib lighter, yellowish green, puberulent; lower surface opaque light grey green, the veins distinct, darker green, glabrous to puberulent, midrib yellow green, subglabrous to puberulent. *Pulvinus* yellow green to olive green, glabrous except for row of hairs along upper side, confluent with petiole, 0.5–9 cm. long in subtending leaf; others 1.5–12(15) cm. long. *Petiole* absent altogether in some populations, frequently absent or reduced in the subtending leaf, green to deep green, glabrous with short line of hairs above to minute sparsely puberulent with shallow groove above in some, 0–24 cm. long in subtending leaf; others 0–56.5 cm. long. *Leaf sheath* chartaceous to semicoriaceous toward base, not auriculate; green or green streaked with purple, purple especially pronounced in lower leaves, the basal portion occasionally entirely dark purple; glabrous to subglabrous, often minute puberulent along margins, appressed-tomentose at base for 1–2 cm. above junction with the stem; 4–26(41) cm. long in subtending leaves, 9.1–120 cm. long in lower leaves. *Stem* green streaked purple or infrequently in basal internodes wholly purple, glabrous except for appressed-tomentose just below the node, and sericeous *ca.* 0.5 cm. above it.

Inflorescence terminal, subtended by a foliage leaf, imbricate, ellipsoid to subcylindric, (2.6)4–9 cm. high and (1.7)2.5–6 cm. in diameter, new shoots commonly arising from the axil of the subtending leaf after flowering. *Peduncle* green or rarely streaked with purple, glabrous below, upper few centimeters minutely puberulent, 4–26.5 cm. long. *Bracts* 4–33, spiral, imbricate, coriaceous, reniform to broadly depressed ovate, subtruncate to rounded, rarely obtuse, 1.9–3 cm. high and 3.1–5.4 cm. wide. Outer surface waxy to semishiny, bright yellow to greenish yellow, basal 1/3–2/3 densely appressed-pilose, upper portion minutely puberulent, apex subglabrous to occasionally glabrous; adaxially, basal portion shiny pale yellowish, glabrous, apical region and upper margins glabrous to puberulent. *Bicarinate prophyll* membranaceous, broadly transverse elliptical, apex rounded; translucent pale yellow to semitransparent, basal portion occasionally with minute purple dots; apical portion puberulent, subglabrous below or subglabrous throughout, inner surface glabrous; 1.5–2.1 cm. high and 1.8–2.9 cm. wide. *Secondary bracts* membranaceous, broadly obovate to transverse broadly obovate, apex rounded to truncate, translucent pale yellow to semitransparent, puberulent, occasionally glabrous throughout, 1.6–1.9 cm. high and 1.65–2.2 cm. wide. *Bracteoles* 2 per flower pair, channeled or carinate, membranaceous, translucent pale yellow to semitransparent, glabrous, 1.5–2.1 cm. high and 0.25–0.5 cm. wide.

Flowers never open spontaneously. *Sepals* slightly stiff, the basal and central portion semispongy, coriaceous, channeled, in bud the apical junction of the three rounded, shiny, basal 1/2 cream to light yellow, pale to bright yellow above, glabrous, 1.55–3 cm. long and (0.4)0.6–1.15 cm. wide. *Corolla tube* cream to pale orange, glabrous, 2.35–3.4 cm. long; corolla lobes inequal, elliptic to broadly elliptic, the median lobe verging on obovate-elliptic, apex obtuse, pale yellow orange to cream, 1.1–2.1 cm. long and 0.43–0.95 cm. wide. *Outer staminode* highly variable, obovate to cuneiform, apex dished, rounded to emarginate, cream to cream white, 0.85–1.5 cm. long and 0.45–0.95 cm. wide, occasionally linear or absent altogether. *Callose staminode* rectangular, frequently bidentate, golden orange to pale orange, the apex occasionally marked with dark red purple, 1–1.3 cm. high. *Cucullate staminode* cream at apex, basal portion golden orange to pale orange, 1–1.2 cm. long and 0.4–0.65 cm. wide, bearing a fingerlike channeled appendage subterminally. *Filament* bearing narrow, lateral appendage to 1.5 cm. wide, uppermost 1 mm. of filament free, appendage attached to lower 1/2 of anther, anther 0.25–0.4 mm. Style and stigma golden. *Ovary* cream white, glabrous, 1.7–3 mm. high. *Capsule* obovoid thin, smooth, straw colored, 1–1.5 cm. high and 0.8–1 cm. in diameter, crowned by a persistent, enlarged calyx. *Seeds* usually 3 per capsule, trigonous, rugose on the outer surface, slate grey, 6–8 mm. high and 4.5–5.5 mm. in diameter, bearing a basal white aril 3–4 mm. high.

Representative specimens seen: ECUADOR. Junction of Guayas, Cañar, Chimborazo, and Bolivar: foothills of the western cordillera near the village of Bucay, 1000–1250 ft., 8–15 June 1945, *W. H. Camp E-3783* (F, GH, K, NY, US). Los Rios: Río Palenque Biological Station, km. 56 Quevedo-Santo Domingo, 150–220 m., 10 April 1972, *C. H. Dodson and M. McMahon 4339* (US); 3 Sept. 1972, *C. H. Dodson and M. McMahon 5062* (US); 220 m., 17 Sept. 1973, *C. H. Dodson and K. Tan 5356* (US).

COLOMBIA. Vaupes: Caño Grande, San José del Guaviare, selva, 240 m., 2 Nov. 1939, *J. Cuatrecasas 7375* (COL). El Valle: Cordillera occidental, vertiente occidental, hoya del Río Dagua, lado izquierdo, Piedra de Moler, bosques, 900–1800 m., 19–28 Aug. 1943, *J. Cuatrecasas 15162* (F, US); cordillera occidental, vertiente occidental, Hoya del Río Sanquininí, lado izquierdo, La Laguna, bosques, 1250–1400 m., 10–20 Dec. 1943, *J. Cuatrecasas 15673* (F); Río Dagua Valley, between La Elsa and Río Blanco, 825 m., 2–5 April 1939, *E. P. Killip 34724* (BM, US); Cisneros, 300–500 m., thickets, 5 May 1939, *E. P. Killip 35545* (US); near Las Ollas, N.W. of Restrepo, 4000 ft., 9 Aug. 1972, *J. W. L. Robinson 193* (COL, K). Santander del Sur: Río Carare, 300–400 m., 28 Sept. 1945, *W. H. Hodge 6509* (GH). Antioquia: en suelos ricos en selvas densas y húmedas en los alrededores de Villa Arteaga, *ca.* 200 m.,

14 Aug. 1948, *J. Araque M.* y *F. A. Barkley 18C748* (COL, US); between Villa Arteaga and Chigorodó, at Lomitas, rain forest, 100 m., 1 Oct. 1961, *J. Cuatrecasas and L. Willard 26109* (COL); cordillera central, selva veranera siempreverde, cerca de Procesito en el valle del Río Medellín, 1100 m., 16 April 1946, *W. H. Hodge 6795* (GH); carretera al mar cerca de Villa Arteaga, ppt. pluv., 3750-5000 mm. ann., 150 m., 4-8 Aug. 1947, *W. H. Hodge 7047* (US); selva pluvial, carretera al mar cerca de Villa Arteaga, ppt., 3750-5000 mm. anuales, 150 m., 16 Dec. 1948, *M. J. Sanchez M.* y *F. López 59* (COL, UC, US). Chocó: Río San Juan, Andagoya, 60-100 m., 15 Aug. 1949, *H. Garcia-Barriga 11278* (COL). Dept. unknown: Chinchiná, 1400 m., June 1952, *M. Koie 5323* (C).

PANAMÁ. Darién: *ca.* 10 mi. S. of El Real on the Río Pirre (House No. 22), 1-11 Aug. 1962, *J. A. Duke 5430* (MO). San Blas: mainland opposite Achituppu, from beach to 1.5 mi. inland, marshy rain forest, 7 Dec. 1966, *W. H. Lewis et al. 122* (GH, MO); *W. H. Lewis et al. 128* (MO). Panamá: along highway near top of Cerro Campana, 10 Sept. 1970, *T. Croat 12083* (SCZ); 2 mi. E. of El Llano, N. of highway in patch of selectively cut forest, in the foothills *ca.* 200 m., 15 Nov. 1970, *H. Kennedy and R. Foster 662* (F); Cerro Campana *ca.* 1 mi. past the Florida State Univ. field station, 800 m., 3 Sept. 1972, *H. Kennedy 1534* (MO). Canal Zone: Mohinga Swamp, lower Chagres River, 26 Aug. 1940, *H. H. Bartlett and T. Lasser 16854* (MICH, MO sterile); drowned forest of Quebrada Culebra, 70-75 m., 19 Dec. 1934, *C. W. Dodge and P. H. Allen 17060* (BM, MO, P); Fort Sherman, 29 Nov. 1969, *J. D. Dwyer 7170* (GH, MO); near Río Frijoles, 9 Sept. 1972, *A. Gentry 5831* (MO); Gatun, Nov. 1859, *S. Hayes 161* (NY), mixed collection: inflorescence *C. marantifolia*, leaf *C. insignis*; *S. Hayes 821* (NY); Orchid Isl., 20 Nov. 1970, *H. Kennedy and R. Foster 689* (DUKE, MO); Pipeline road near Río Agua Salud, N.W. of Gamboa, 21 Oct. 1972, *H. Kennedy and J. Redemske-Young 1828* (MO); lake shore along Gatun River Valley, 16 Aug. 1914, *H. Pittier 6746* (US); Cuatro Calles Hills, near Matachín 20-150 m., 23 July 1911, *H. Pittier 4050* (US); drowned forest of Quebrada Ancha, 70 m., 21 Dec. 1934, *J. A. Steyermark and P. H. Allen 17060* (MO); 6 mi. N. of Gamboa near Río Frijol, 6 Oct. 1965, *E. L. Tyson 1436* (MO); Gatun, Jan. 1858, *M. Wagner s.n.* (M). Barro Colorado Island: 1931, *S. Aviles 15* (F); Zetek trail 500, 17 Sept. 1970, *T. Croat 12222* (F); Gatun Lake, along stream in forest, 17 Jan. 1924, *P. C. Standley 31392* (US); wet forest, 120 m., 18-24 Nov. 1925, *P. C. Standley 40927* (US); along stream west of Donato Trail, forest, 100-200 m., 23 Nov. 1948, *E. P. Killip 39978* (US). Colón: 2 mi. N.E. from Puerto Pilón on trail along Río Piedras, 7 Sept. 1967, *M. D. Correa A. and B. L. Haines 247* (MO); 6 mi. S. of Portobelo, 17 July 1970, *T. Croat 11410* (F); bridge over Río Buenaventura, semi-inundated forest and tierra firma, near Portobelo, 27 Aug. 1970, *H. Kennedy 479* (NY); trail behind Peluca meteorological station, old secondary forest, 25 March 1973, *H. Kennedy and R. L. Dressler 2969* (F), sterile; near Peluca, along the Río Boqueron near the road to Nombre de Dios, 17 Sept. 1973, *H. Kennedy and R. L. Dressler 3302* (SCZ). Coclé: near La Mesa, El Valle de Antón, cloud forest, *ca.* 800 m., 2 Aug. 1970, *H. Kennedy and J. Luteyn 447* (DAV, DUKE, MO); along road 8 mi. N. of El Valle de Antón, follow road to end in farmer's field (La Mesa) and then to right, 3 Aug. 1970, *J. L. Luteyn and H. Kennedy 1716* (DUKE). Bocas del Toro: Almirante, along a stream, S.W. of town, 18 Nov. 1971, *H. Kennedy and R. L. Dressler 1285* (B, BM, COL, DAV, DUKE, F, GH, MO, PMA, S, U, US, VEN); 10-15 mi. inland (S.) from mouth of Changuinola River, banks of the river and adjacent rain forest, 18 Dec. 1966, *W. H. Lewis et al. 983* (GH, MO, US); Water Valley, 14 Sept. 1940, *H. von Wedel 772* (BM, GH, US); Isla Colón, 23 Oct. 1940, *H. von Wedel 1328* (GH, MO, US).

COSTA RICA. Limón: banana and cacao plantation on level areas between Siquirres and the Río Pacuare, and remnant forest on steep hills S. of the railroad bridge over the Río Pacuare at 50-100 m., 10° 5′ N. x 83° 29′ W., 20-22 Dec. 1969, *W. C. Burger and R. L. Liesner 6929* (F, GH, S); *W. C. Burger and R. L. Liesner 6954* (F); near Limón, Caribbean coastal plain with cacao plantation, remnants of native vegetation, wet forest vegetation and littoral vegetation between sea level and 100 m., in the general vicinity of Limón, 3-7 Sept. 1971, *W. C. Burger and M. Burger 8465* (F); palm forest W. of main tramline to Golden Grove, 30 m., river bank, 28 July 1936, *C. W. Dodge and V. F. Goerger 9469* (MO); forest near Finca Castilla, 30 m., 29 July 1936, *C. W. Dodge and V. F. Goerger 9497* (MO); forest near farmhouse, Finca Castilla, 30 m., 29 July 1936, *C. W. Dodge and V. F. Goerger 9498* (MO); vicinity of U.S. Dept. of Agri. Rubber Exp. Station, Los Diamantes on Río Santa Clara (1.6 km. E. of Guapiles), 11 July 1949, 200 m., *R. W. Holm and H. H. Iltis 392* (GH); road between mile 28 and Zent, *ca.* 3/4 way to Zent, 27 Feb. 1972, *H. Kennedy, R. Andrews, R. L. Dressler 1381* (CR, GH, MO, US); La Lola Cacao Institute, near mile 28, within cacao plantation and adjacent secondary forest, 29 Feb. 1972, *H. Kennedy and R. L. Dressler 1389* (COL, F, US); Siquirres, 6 Dec. 1931, *W. Kupper 142* (M); Matina, Aug. 1895,

H. Pittier 9765 (BR, US); Guapiles (Pococí) Finca "La Granja," 1936, *F. Solís R. 418* (CR, F); Bananales de Zent, 31 m., Sept. 1901, *A. O. Tonduz 14547* (CR); Hacienda de Zent, *United Fruit 320* (US). Puntarenas: swamps near Jalaca station, region of Golfo Dulce and Palmar Norte, *ca.* 0 m., 23 July 1951, *P. H. Allen 6277* (F); area between the Estero Azul and the Río Sierpe, sea level, 20 Oct. 1952, *P. H. Allen 6601* (F, GH, US); forested slopes east of Las Cruces, 5-6 km. S. of San Vito on and around the property of Mr. Robert Wilson, 1100-1200 m., 8° 47′ N. x 82° 58′ W., 15-16 Jan. 1967, *W. Burger and G. Matta U. 4407* (CR, F); flood plain of Río Sandalo, peninsula of Osa, 1-10 m., 21 Aug. 1936, *C. W. Dodge and V. F. Goerger 9954* (MO); 22 Aug. 1936, *C. W. Dodge and V. F. Goerger 10019* (BM, CR, F, MO); 24 Aug. 1936, *C. W. Dodge and V. F. Goerger 10145* (GH); 2.7 km. E.S.E. of Monteverde, 1540 m., 1 Oct. 1970, *H. Kennedy and G. Powell 577* (F); old forest above Wilson's Finca, 6 km. S. of San Vito de Java, *ca.* 5000 ft., 14 Aug. 1967, *P. Raven 21754* (CR, F). San José: along road from Santiago to Parrita, 700 m., 24 Sept. 1967, *R. Lent 1328* (F). Cartago: Turrialba, forest below the Agricultural Institute, *ca.* 700 m., 12 Oct. 1970, *H. Kennedy 641* (CR, F); Turrialba Inst. forest, 600 m., 30 Sept. 1953, *J. León 4206* (USJ); in vicinity of the Inst. Interamericano de Ciencias Agricolas in Turrialba, in secondary shady forest on hillsides, 725 m., 6 Aug. 1969, *J. van Rooden and L. M. A. Akkermans 113* (U); Haies à Turrialba, 570 m., Nov. 1893, *A. O. Tonduz 8310* (BR, CR), this collection was cited by Schumann (1902) under *C. macrosepala* along with two other taxa and is excluded by the lectotypification of *C. macrosepala*; Broussailles, lieuse frais à Las Vueltas, Tucurrique, 635 m., Mar. 1899, *A. Tonduz 13300* (US). Alejuela: Quebrada Lajas, Finca Los Ensayos, Buena Vista de San Carlos, 850 m., 10 Sept. 1964, *A. Jiménez M. 2322* (F); 3.5 km. E.S.E. of Monteverde, 1460 m., 30 Sept. 1970, *H. Kennedy and G. Powell 576* (CR, F, US); premontane forest, 15-20 km. N.N.W. of San Ramón along Calle Los Angeles, beyond La Balsa, *ca.* 1000 m., 19 Sept. 1972, *H. Kennedy 1666* (CR, DAV, F, GH, MO, US); Heredia: Puerto Viejo, Finca La Selva, second growth, 90 m., 16 Sept. 1969, *G. W. Frankie 393C* (F); La Selva O.T.S. field station on Río Puerto Viejo, near Pto. Viejo, 6 Aug. 1968, *H. Kennedy 60* (F); 15 July 1970, *H. Kennedy 407* (F, US). Guanacaste: El Silencio near Tilarán, *ca.* 750 m., 13 Jan. 1926, *P. C. Standley and J. Valerio 44833* (US).

NICARAGUA. Zelaya: vicinity of El Recreo near sea level, 7 Aug. 1947, *L. E. Long 66* (F). Dept. unknown: near Tortuga, May 1847, *A. S. Ørsted s.n.* (C) mixed collection, part *C. marantifolia* part *C. donnell-smithii.*

HONDURAS. Atlantida: Lancetilla Valley, 22 June-27 July 1929, *A. M. Chickering 225* (MICH); near Tela, Lancetilla Valley, wet mt. forest, 20-600 m., 6 Dec.-20 Mar. 1928, *P. C. Standley 52737* (F, US), *54316* (US), *54928* (F, US); near Lancetilla in wet marshy situation, 90 ft., 5 Aug. 1934, *T. G. Yuncker 4917* (F, MICH, MO); on the mountain slopes and coastal plains, vicinity of La Ceiba, in thicket, lower slopes of Mt. Cangrejal, 16 July 1938, *T. C. Yuncker, J. M. Koepper and K. A. Wagner 8460* (F). Cortés: bosque lluvioso de la Montaña La Zona, cañon del Río Piedras 3 km. de San Pedro Sula, 190 m., 28 Nov. 1950, *A. Molina R. 3400* (F). Dept. unknown: Puerto Siena, forest along trail near Eliott's plantation, 19 Jan. 1903, *P. Wilson 100* (NY, US).

GUATEMALA. Izabal: between Los Amates and Izabal, Sierra del Mico, 1000 ft., 26 Feb. 1908, *W. A. Kellerman 7512* (US); Los Amates, 90 m., 15 Feb. 1908, *W. A. Kellerman 7318* (US); ruinen von Quiriguá in feuchtem Urwald, 2 Jan. 1897, *E. Seler 3320* (GH, NY, US). Escuintla: between Río Jute and Río Panaleon on road between Escuintla and Santa Lucía Cotz, 24 Jan. 1939, *P. C. Standley 63549* (F). Petén: matorrales y bosque húmedo orilla Río Pasión camino a Seibal, 18 Nov. 1965, 60 m., *A. Molina R. 15839* (F). Dept. unknown: Bananera, 10 Sept. 1936, *Mrs. B. B. Lewis 279* (F).

Flowering: *Calathea marantifolia* flowers mainly in the rainy season, July through December. In Guatemala it was found flowering in February and in El Valle, Panamá, in April.

Pollinator: Because *Calathea marantifolia* is wide ranging, occurring in wet to semideciduous forest habitats, it is visited by more pollinator species than are other closed-flowered Calatheas. Populations at Las Cruces (near San Vito de Java), La Selva (near Puerto Viejo), and Turrialba in Costa Rica and at Cerro Campana in Panamá were observed to be pollinated by *Eulaema speciosa* (Mocsary). Those in the open garden situation at San Vito were occasionally pollinated by *Eulaema cingulata* (Fabricius). *Eulaema cingulata* was observed to be the main pollinator for those on Pipeline Road, Canal Zone. In a large, dense population in an exposed meadow near Monteverde (Costa Rica) both *E. cingulata* and *E. polychroma* (Mocsary) were equally important as pollinators. Near Portobelo on the Río Buenaventura (Panamá) both *Euglossa ignita* Smith and *Eulaema speciosa* were observed to pollinate, while on Barro Colorado Island pollination was predominantly by *Euglossa imperialis* Cockerell.

Distribution and habitat: *Calathea marantifolia* has a wide distribution, extending from Ecuador through Colombia and Panamá, northward into the Department of Escuintla in Guatemala. In Central America this species occurs on both the Atlantic and Pacific slopes. At the dry end of its range, *C. marantifolia* occurs in seasonal, semideciduous forests, whereas at the wet end of the range it occurs in the rain forests of Chocó, Colombia. They are found predominately in the evergreen (wet) forest and premontane wet to rain forest habitats. In the semideciduous forests they occur within the forest and often on stream banks or near streams. In contrast, plants of *C. latifolia* growing sympatrically with it on Pipeline Road, Canal Zone, occurred in the open, but not in the forest, whereas *C. marantifolia* occurred in the forest. In the wet and premontane rain forest they often occur near a forest edge but usually within the forest, such as along the roads in Cerro Campana, Panamá, and San Vito, Costa Rica, or around the perimeter of tree fall areas. In wetter areas such as around Portobelo and Bocas del Toro, Panamá, and Monteverde, Costa Rica, they may occur in open stands, especially if the soil is quite moist.

Variation: This species is variable throughout its range. The Atlantic coastal populations were described as a distinct species, *Calathea lagunae,* by Woodson (Woodson and Schery, 1942). Plants from Ecuador have a densely velvety tomentose lower leaf surface; the sheath, petiole, and peduncle are densely short puberulent. Labels on the Ecuadorian collection (*Camp E-3783*) and the Colombian collections from El Valle (*Cuatrecasas 15162, 15673*), and Antioquia (*Araque and Barkley 18C748*; *Hodge 7047*; *Sánchez and López 59*) all make mention of purple sepals or fruit, which is unusual for this species. I have not seen living material from these areas but herbarium specimens agree in other respects with *Calathea marantifolia*, and they might be merely a color form of it. However, they all have in common a densely tomentose underleaf, puberulent leaf sheath, and an apparently thicker-textured leaf than Central American plants, perhaps constituting a recognizable variety were more collections known. The pubescence of the underleaf varies as much between nearby populations as it does between distant ones, though plants from Monteverde, Costa Rica, and the Atlantic coast in Bocas del Toro Province, Panamá, and Limón Province, Costa Rica, are subglabrous with the veins minutely puberulent below. In all populations the pubescence on the bract is very dense at the base and is composed of longer hairs than those on the upper part of the bract. In populations near Portobelo, Panamá, the pubescence at the base of the bract is subvillous rather than tomentose. The apex of the bicarinate prophyll and secondary bracts of most plants are pubescent. However, in the Monteverde population they are glabrous, and in plants from Bocas del Toro and Limón Provinces they are subglabrous. Likewise, the adaxial surface of the bract apically is puberulent in most populations, but is glabrous in individuals from the mid elevations of Cerro Campana and El Valle, Panamá, and Monteverde, Costa Rica. The Monteverde population differs from the rest in several features; the bracts are much less pubescent, having a shiny, waxy appearance (more like *C. trichoneura*), the apex of the bract is obtuse rather than rounded, and the petiole is absent, the sheath adjacent to the pulvinus. Plants intermediate between this form and those from San Vito, Costa Rica, were found near San Ramón, Costa Rica. Plants from the Atlantic coast around Bocas del Toro, Panamá, and Limón to Puerto Viejo in Costa Rica are characterized by smaller, usually more numerous leaves, fewer bracts, and a more slender overall habit. In addition, the corolla tube in the Atlantic coast individuals is *ca.* 2-2.6 cm. long, whereas those of plants near San Vito, Costa Rica, are 2.6-3.4 cm. long; the sepals also are in the lower range, *ca.* 1.5-1.8, opposed to an average of 2-3 cm. long in San Vito plants. In the Bocas del Toro populations, some plants have much paler petals and sepals than others in the population: the petals are nearly cream white instead of the usual cream yellow or light yellow, and the sepals are a faint yellowish. Plants from Puerto Viejo are all of the pale color morph. Within the range of *C. marantifolia,* populations are found to be slightly divergent one from the other, those from South America, Ecuador, and Colombia represent one group; the plants from the Darién, Cerro Campana, El Valle, and the Canal Zone in Panamá, San Vito and Turrialba in Costa Rica, and Lancetilla Valley, Honduras, represent another fairly uniform group; the Atlantic coastal populations of Bocas del Toro and Limón Provinces another; and the Monteverde, Costa Rica, population a fourth. Flowering individuals in the open are smaller than those at the forest edge and within the forest, often considerably so. Smaller individuals tend to have proportionally longer, narrower leaves than larger ones. The individual cited as type by Standley (1927) is an unusually small specimen, but similar small individuals have been seen both at Monteverde and in Bocas del Toro. Other specimens collected by Standley and Valerio near Tilaran (*Standley and Valerio 44833,* US) are larger but similar in qualitative features.

Relationship: Calathea marantifolia is most closely related to *C. trichoneura,* sharing a common habit, leaf morphology and texture, and color and texture of the bracts and sepals. In *C. trichoneura* the bracts have a shiny almost waxy sheen, and the bract apex is obtuse, both features that are shared by the population of *C. marantifolia* near Monteverde, Costa Rica, though these lack the pubescent veins. *Calathea marantifolia*, however, lacks the color change with age. It is more distantly related to *C. donnell-smithii*, sharing a similar leaf morphology and similar purple striping of the leaf sheath and stem, though differing in habit and color of bracts and sepals. Though apparently quite distantly related to *C. warscewiczii*, a hybrid between the two species was found near La Lola, Limón Province, Costa Rica. In habit, inflorescence morphology, and the morphology and texture of the flower, *C. marantifolia* bears a close resemblance to *C. latifolia*.

Notes: Nine flowers of *Calathea marantifolia* on Pipeline Road, Canal Zone, were artificially self-pollinated. All nine flowers set seed indicating that this species is probably self-compatible. Of thirgy-eight crosses attempted with *Calathea latifolia* at Pipeline Road, Canal Zone, none was successful, while both species set seed when selfed. However, some plants of *C. marantifolia* between Coco Solo Hospital and Four Corners, Canal Zone, showed distinct purple bands on the underleaf of juvenile plants. The possibility that these are of hybrid origin between *C. marantifolia* and *C. latifolia* seems remote, but a hybrid between *C. marantifolia* and a less closely related species than *C. latifolia* has been observed. The seeds germinate at the end of the rainy season, and the new plants pass the dry season as seedlings.

Petersen (1890) saw specimens from "Gatum in Panama" (*Wagner s.n.*) and from "Guyane" (*Leprieur s.n.*), which he misidentified as *Calathea violacea* Lindley, a species of southern Brasil, and described *C. violacea* var. *hirsuta* Petersen. The Wagner specimen in Munich is annotated "*Calathea violacea* Lindl. var. *hirsuta*." The handwriting appears to be that of Petersen, although Eichler's is so similar that it may be his. However, Petersen observed (p. 114), "In specimine Wagneriano caulis innovatione ex axilla folii caulini ramificatus est." The Munich specimen clearly fits this description; a shoot is present in the axil of the subtending leaf of the inflorescence. The bracts of the Wagner specimen and those of the Leprieur specimen are pubescent, though the Wagner plant more closely approaches the current definition of hirsute. However, it is not known if this definition (*sensu* Lawrence, 1951, p. 755) is that which Petersen intended. The Wagner specimen was chosen as lectotype because it is clear Petersen saw this material, and further, the other cited specimen, that of Leprieur, was annotated by Petersen as "*Calathea violacea* var. *hirsuta*?" showing that he himself entertained some doubt as to its identity. It is in fact a different species than the Wagner specimen and is most closely allied to *C. truncata*, or possibly conspecific. Neither the Leprieur nor the Wagner specimen is referrable to *Calathea violacea* of Lindley.

5. **Calathea trichoneura** H. Kennedy, sp. nov.

(Plate 9, Figure 12)

Planta robusta. Folia longe petiolata, petioli pars superior callosa teres glabra super puberula, ovata obtusa vel rotundata cum acumine base rotundata vel truncata supra smaragdina venas pubescentia ceterum glabra medianum puberula pallide viridia, subtus opaca pallide viridi-grisea subglabra; vagina subcoriacea, viridis vel vittata purpurea glabra vel minutissime puberula. Spica imbricata ovoidea, pedunculo viridi glabro; bracteae spiraliter dispositae coriaceae, initio late obtrullatae vel depresse ovatae obtusae vel rotundatae, demum rombicae acutae aureae postea rubrae apice patenti glabrae raro base puberulae; paria florum 5 vel ultra bracteolis membranaceis, prophyllis et mesophyllis flavis glabris; ovarium album glabrum; sepala eburnea supra lutea postea rubra; corollae tubus albus supra eburnea, lobi eburnei supra pallide lutacei late ovati vel late subobovati obtusi; staminodium exterius cremeum obovatum rotundatum vel bilobatum; callosum rectangulare truncatum vel bidentatum; cucullatum cremeum. Capsula turbinata obovata glabra straminea supra nebulose rubra calyce demum coronata; semina trigona dorso rugosa rotundata schistacea, arillo albo munita.

Herb to 2.5 m. high. Each mature shoot bearing 3-4 leaves, 0-1 basal, 2-3 cauline. Roots bearing root-tubers, especially in younger plants. *Cataphylls* thick, semispongy at base, chartaceous above, narrowly ovate, apex acute, mucronulate, green finely streaked purple giving a greenish purple color, lower ones almost solid purple, minutely puberulent; within, shiny pale green to whitish, innermost 81-104 cm. high, others 18.1-57.2 cm. high. *Leaf blade* ovate, apex rounded with a point, leaf base rounded to truncate, at junction with pulvinus, shortly acuminate, leaf subtending the inflorescence 25-53.3 cm. long and 14.3-27.3 cm. wide; others 34-68.6 cm. long and 18-32 cm. wide. Leaf surface above shiny in young leaves, dulling with age, green, midrib yellow green, major veins raised, tomentose with tan hairs, margins and center of midrib tomentose; lower surface dull grey green subglabrous to minutely puberulent along the nerves. *Pulvinus* olive green, glabrous except line of hairs above, 0.9-7 cm. long. *Petiole* green,

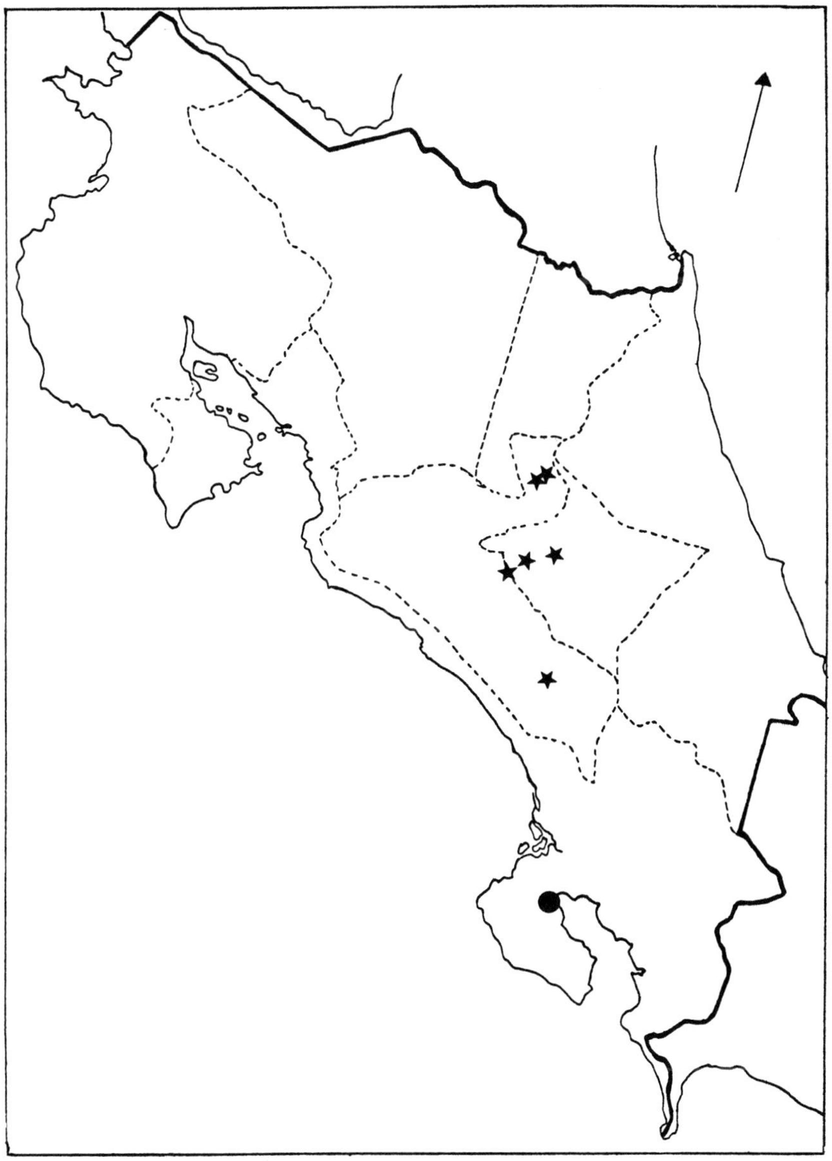

Fig. 12. Distribution of *Calathea longiflora* (circles) and *C. trichoneura* (stars) in Costa Rica.

glabrous, 0–13.3 cm. long, distinctly grooved above, groove *ca.* 2 mm. wide. *Leaf sheath* subcoriaceous to papyraceous, green streaked purple at base especially in lower leaves, glabrous to minutely puberulent especially in the lower 1/2–2/3 and on the margins, 9.5–26.1 cm. long in subtending leaf, others 20–163 cm. long. *Stem* green, just above nodes may be finely streaked purple, shiny, glabrous.

Inflorescence terminal, subtended by a foliage leaf, imbricate, ovoid, 5–9 cm. high and 4–6.3 cm. in diameter. *Peduncle* green, glabrous, puberulent at the apex, 14–28.3 cm. long. *Bracts* 5–13, spirally arranged, imbricate, coriaceous, margin patent to slightly recurved, lower bracts broadly obtrullate (1:1

ratio) to approaching depressed-ovate, apex obtuse to rounded, abruptly acuminate centrally, the first bract usually caudate-acuminate with acumen 6–7 mm. long; uppermost bracts rhombic, apex obtuse to acute. Throughout the inflorescence the trend is from broad short bracts at the base to taller narrower ones toward the apex. Lower bracts 2.5–3.1 cm. high (including caudate tip on first bract up to 3.3 cm. high) and 3.1–4.5 cm. wide; upper bracts up to 3.5 cm. high and 2.6 cm. wide. Abaxial surface bright yellow turning red with age, semiwaxy in appearance, glabrous, rarely puberulent above, puberulent at junction with peduncle or rachis; adaxial surface yellow, glabrous, recurved portion turning red with age. In old inflorescences the bracts are entirely red. Each bract subtends at least 5 flower pairs. *Bicarinate prophyll* membranaceous, circular to approaching broadly obovate, apex rounded, translucent yellow, glabrous, with age apex turning red, 1.8–2.2 cm. high and 1.9–2.2 cm. wide. *Secondary bracts* membranaceous, broadly oblong to subquadrate, truncate, occasionally in upper bracts with a short acumen, translucent yellow, with age apex turns red, glabrous, 1.8–2 cm. high and 1.5–1.9 cm. wide.

Flowers never open spontaneously. *Sepals* channeled, base and central portion semispongy narrowly obovate, apex obtuse, shiny, pale cream white at base, yellow above, glabrous, 3.3–4.1 cm. long and 0.9–1.2 cm. wide. *Corolla tube* cream white, glabrous 3.4–4.1 cm. long; corolla lobes subequal, broadly elliptical to broadly subobovate, apex obtuse, margins inrolled appearing acute, cream at base, upper 2/3 light yellowish, 1.8–2 cm. long and 0.9–1.1 cm. wide. *Outer staminode* obovate, rounded to irregular or shallowly emarginate, cream, 1.4–1.5 cm. high and 0.6–0.7 cm. wide. *Callose staminode* rectangular, truncate, bidentate centrally, pale cream, 1.3–1.4 cm. high. *Cucullate staminode* cream, 1.1–1.2 cm. long and 0.6–0.7 cm. wide, provided with a narrow channeled appendage subterminally to 2 mm. long. *Filament* of stamen with lateral narrow petaloid appendage *ca.* 1.5 mm. wide; anther free in apical 1/4, 4–5 mm. long. Style and stigma golden. Ovary white, glabrous, *ca.* 3 mm. high. *Capsule* obovoid, yellow tinged red to red purple apically, *ca.* 1.1 cm. high and *ca.* 0.8 cm. in diameter, crowned by a persistent live expanded calyx, sepals turning red with age apically. *Seeds* usually 3 per capsule, slate grey, trigonous, rugose on outer surface, 7–8 mm. high and 5.5–6 mm. wide, provided with a basal white aril *ca.* 3 mm. long.

Type: COSTA RICA. Cartago: *ca.* 10 km. S. of Tapantí along the road above Río Grande de Orosi, cloud forest, *ca.* 1500 m., 7 Sept. 1972, *H. Kennedy and L. D. Gómez P. 1571* (holotype F!; isotypes CR!, DAV!, K!, MO!, U!, US!).

Specimens seen: COSTA RICA. Cartago: steep slopes with open forest and many epiphytes about 15 km. S. of Tapantí along new road, on the east slope above the Río Grande de Orosi near the concrete bridge at 1500 m., 9° 47′ N. x 83° 47′ W., 12, 14, 17 Dec. 1969, *W. C. Burger and R. L. Liesner 6828* (F); about 10 km. S. of Tapantí along the new road on the east slope above the Río Grande de Orosi, *ca.* 1600 m., 9° 42′ N. x 83° 46′ W., 14 and 17 July, *W. Burger and M. Burger 7587* (F); 1 Sept. 1971, *W. C. Burger and M. Burger 8427* (F); *ca.* 10 km. S. of Tapantí along the road above Río Grande de Orosi, cloud forest, *ca.* 1500 m., 10 Sept., 1971, *H. Kennedy and W. Ramírez 1144* (US); primary forest near Río dos Amigos, 8 km. S. of Tapantí, 1600 m., 17 Oct. 1971, *R. W. Lent 2202* (F); La Estrella, wet forest, 26–27 Mar. 1924, *P. C. Standley 39500* (US); El Muñeco on the Río Navarro, 1400–1500 m., 6–7 Mar. 1926, *P. C. Standley and R. Torres R. 51367* (US); *51394* (US); *51706* (S, US). San José: steep wet slopes and forested stream sides, Río Claro Valley at 1000–1200 m., (Río La Hondura drainage) below La Palma, N.E. of San Jerónimo, 10° 3′ N. x 83° 58′ W., 19 Nov. 1969, *W. C. Burger and R. Liesner 6244* (CR, S); along Pan Am Highway about halfway between San Isidro del General and Cerro de La Muerte, *ca.* 1400 m., 26 Aug. 1968, *H. Kennedy and R. B. Foster 74* (US); below Bajo La Hondura, Río La Hondura drainage, steep slopes, 1100 m., 20 Sept. 1972, *H. Kennedy 1686* (MO); trail beyond Bajo La Hondura towards Río Claro, cloud forest at approx. 1100–1200 m., 4 July 1972, *J. L. Luteyn 3321* (DUKE); Bords des eaux à La Palma, 1459 m., 8 Sept. 1898, *A. Tonduz 12540* (US).

Flowering: *Calathea trichoneura* flowers mainly during the rainy season and is represented in collections from March through December. It would not be surprising to find it in flower in January and February as the soil moisture is moderately high year round in the area where it is found.

Distribution and habitat: Costa Rica. *Calathea trichoneura* occurs between 1000 and 1600 m. from premontane rain forest to lower montane rain forest (cloud forest), on the Pacific slope of the Cordillera de Talamanca, on the slopes (Atlantic drainage) of the Cordillera Central and the mountain range to the southeast of San José. It mostly occurs along stream sides in moderate shade, very often close to the water's edge.

Variation: The most noticeable differences between the plants of *Calathea trichoneura* from Tapantí and those from Bajo La Hondura are the nature of the pubescence and shape of the bracts. The former plants are glabrous, the apex of the first bract is usually caudate, and the uppermost bracts acute, whereas in the Bajo La Hondura plants the bracts are appressed-pilose below, puberulent above, the apex of the first bract bears a short acuminate point, and the uppermost bracts are obtuse. The pubescence on the veins is less pronounced in the Bajo La Hondura plants than in the Tapantí or Pacific slope ones, and the last two have a somewhat narrower leaf. The above characters are similar to those of *Calathea marantifolia*, the most closely related species, and might be the result of a past incidence of introgression with that species.

Relationship: *Calathea trichoneura* is most closely related to *Calathea marantifolia*. Both share a similar habit, leaf morphology and color, and texture of bracts and flowers. The population of *C. marantifolia* near Monteverde, Costa Rica, has a more pointed (obtuse) bract apex, similar to that of *C. trichoneura*. *C. marantifolia*, however, never exhibits a bract color change as in *C. trichoneura*.

Notes: Individuals from Tapantí, Costa Rica, grown in cultivation in the Canal Zone near Ancon did poorly and produced virtually no pollen; transference of the few grains produced into the stigma produced no seed. In Tapantí many inflorescences had maturing capsules and the flowers produced normal amounts of pollen.

In the original description of *Calathea macrosepala*, Schumann cited *A. Tonduz 12540* (US) among his (then) syntypes. This specimen differs from the original description in possessing puberulent veins. It differs from the Guatemalan collections also cited in that it is found at moderately high (1000-1600 m.) elevations and does not die back to the rhizome during the dry season.

6. *Calathea vinosa* H. Kennedy

(Plate 10, Figure 13)

Calathea vinosa H. Kennedy, Ann. Missouri Bot. Gard. 60: 422-424. 1973. Type: Cultivated at Finca Las Cruces, property of Robt. Wilson, Las Cruces, 5-6 km. S. of San Vito from rhizomes collected in the forest N.E. of the Tropical Science Center field station, *ca.* 50 m., 5 km. W. of Rincón de Osa, 8° 42′ N. x 83° 31′ W., Prov. Puntarenas, Costa Rica, 23 Sept. 1969, *H. Kennedy 384* (F!; isotypes COL!, MO!, US!).

Caulescent herb 1.3-2.3 m. high. Shoots bearing 2-4 leaves; 1-2 cauline, 1-2 basal. *Cataphylls* coriaceous, thicker and somewhat fleshy toward the base, ovate to narrowly ovate-triangular, apex acute to obtuse, mucronulate, abaxially red purple, subglabrous to finely tomentose (both straight and retrorsely barbed hairs) at the apex and margins; adaxially light green and glabrous; upper portions rotting away with age, innermost cataphyll 60-93 cm. long. *Leaf blade* oblong to linear-oblong, occasionally narrowly ovate in smaller leaves, apex obtuse to rounded with a point, base rounded to truncate, junction of blade to pulvinus shortly acuminate; coriaceous, stiff rather than pliable, 65-105 cm. long and 17.5-38.7 cm. wide. Leaf blade above dark semimetallic bluish green, glabrous, midrib light yellow green, puberulent; leaf surface below dark red purple, densely tomentose with retrorsely barbed hairs, each hair arising from a swollen bulbous base. *Pulvinus* terete, dark red purple commonly concolorous, or petiole occasionally slightly darker or lighter, glabrous except for row of hairs above, confluent with petiole and virtually of same diameter, (3)5.5-15 cm. long. *Petiole* red purple, glabrous to subglabrous with a few scattered hairs along upper surface, occasionally bearing a faint, shallow groove to 1.5 mm. wide on the upper side, 0.9-29 cm. long in subtending leaf, others to 59 cm. long. *Leaf sheath* not auriculate, dark red purple, centrally glabrous with the clasping margins moderately short pubescent with straight hairs becoming especially dense at the apex, stiff semibrittle; whitish green, glabrous within; sheath of subtending leaf 17-31.5 cm. long, others 48-144 cm. long. Base of the leaf sheath and upper 2-3 cm. of the internode puberulent.

Inflorescence terminal, subtended by a foliage leaf, subimbricate, broadly cylindrical to slightly ovoid, 12-14(16.8) cm. high and 7-8.5 cm. in diameter. *Peduncle* dark red purple, minutely tomentose with straight hairs, 10-19 cm. long. *Bracts* spiral, first few reniform, apex obtuse, later bracts longer and narrower, broadly trullate-ovate to trullate-ovate, apex acute to acuminate, often slightly eccentric, outer margin and apex recurved, occasionally patent in young inflorescences, coriaceous, 3.6-5.5 cm. high and 4.3-6.6 cm. broad. Basal portion of bract pale straw colored, recurved apical margin ivory or tinged

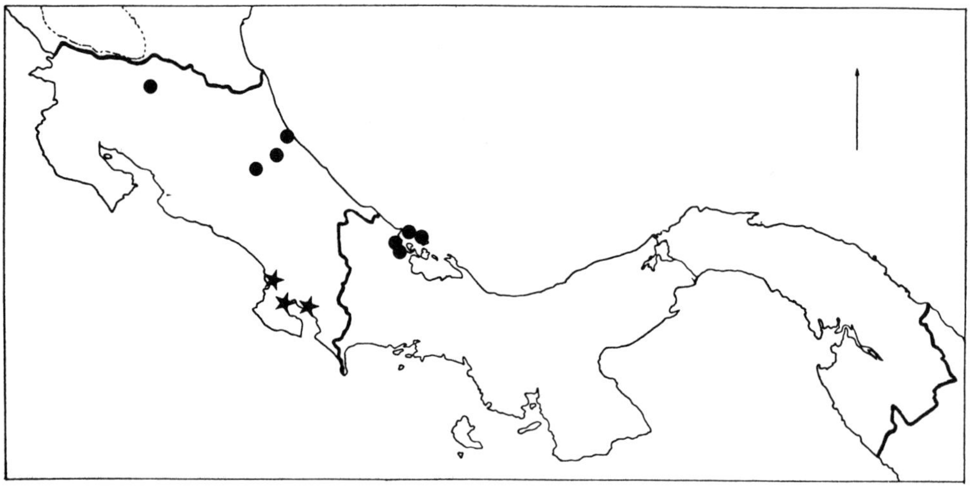

Fig. 13. Distribution of *Calathea vinosa* (stars) and *C. indecora* (circles) in Costa Rica and Panamá.

greenish; upper bracts often bearing slight streaks of purple adaxially at the apex, outer surface and recurved margin densely velvety tomentose, basal half of inner surface glabrous. Each bract subtending to 6 flower pairs. *Bicarinate prophyll* membranaceous, broadly transverse elliptic, obtuse, translucent pale yellowish green, densely puberulent, glabrous within except apex and margins, (2.5)2.6–2.9(3.1) cm. high. *Secondary bracts* membranaceous, broadly to transverse broadly obovate, truncate to retuse, translucent, white, subglabrous at base, upper 1/2 pale yellow green, densely puberulent, *ca.* 2.5–2.6 cm. high and 3–3.1 cm. wide. *Bracteoles* subtending individual flowers, 2 per flower pair, membranaceous, carinate, elliptic to obovate, obtuse, translucent pale yellow green, puberulent at apex, *ca.* 2.5–2.6 cm. high and *ca.* 1–1.2 cm. wide.

Flowers never open spontaneously. *Sepals* channeled, oblong to narrowly obovate, obtuse, apical margins infolded, appearing acute, pale yellow, glabrous, 3.55–3.9 cm. long and (0.9)1.05–1.3(1.5) cm. wide, about 0.5 cm. wide at base. *Corolla tube* white at base, cream colored above, glabrous, 4.2–4.5 cm. long; petals subequal, oblong to subobovate, obtuse, cream colored, glabrous, 1.6–1.75 cm. long and 0.8–1.1 cm. wide. *Outer staminode* narrowly obovate to obtriangular, bifid, apical 1/2 shallowly cupped, white, 1.1–1.3 cm. long and 6–7.5 mm. wide. *Callose staminode* rectangular, apex truncate, irregular, with a small point in the center, completely callose, pale golden, 1.15–1.25 cm. high. *Cucullate staminode* cream colored, 1.1–1.2 cm. long and 0.5 cm. wide, provided with a subterminal filiform, channeled appendage to 1.5 mm. high. *Stamen* with lateral petaloid appendage to 1.5 mm. wide extending to the tip of the anther; anther 4.5 mm. long. Style and stigma golden. *Ovary* white, glabrous, 3.5–4.5 mm. high and to 2.5 mm. in diameter. *Capsule* thin, obovoid pale cream yellow, glabrous, *ca.* 1.7 cm. high and *ca.* 0.9 cm. wide, crowned by a persistent, enlarged calyx; sepals up to 1.45 cm. wide at base. *Seeds* usually 3 per capsule, trigonous, rugose on outer surface, slate grey, 9–9.5 mm. high and 4.5–5 mm. in diameter, bearing a basal white aril to 3.5 mm. high.

Specimens seen: COSTA RICA. Puntarenas: vicinity of Grande de Terraba, sea level, 1 Nov. 1949, *P. H. Allen 5340* (US); entre Palmar Norte y Puerto Cortéz, 50 m., 6 Aug. 1964, *A. Jiménez M. 2240* (F); Peninsula de Osa, 8 km. al Sur de Rincón, en foresta cenagosa (swamp forest), 28 Feb. 1965, *A. Jiménez M. 3024* (F, CR); Golfito, watershed forest, *ca.* 20 m., 17 Oct. 1970, *H. Kennedy 644,* (DAV); Osa Peninsula, 2.5 mi. S.W. of Rincón, near T. S. C. field station (airfield), 15–30 m., flowered in cultivation at Las Cruces near San Vito, 20 Oct. 1970, *H. Kennedy 651* (F); Osa Peninsula, 2.5 mi. S.W. of Rincón, across the Río Agua Buena from T. S. C. field station, 15–30 m., 8° 42′ N. x 83° 29′ W., 10 Sept. 1972, *H. Kennedy 1616* (DUKE, MO, US).

Flowering: *Calathea vinosa* flowers in the mid to late rainy season, August to December. One flowering specimen was collected as late as 28 February, but none was found in flower in early April nor were remnants of old inflorescences seen then.

Pollinator: In both the type locality (on the Osa Peninsula) and in Golfito, *Calathea vinosa* was observed to be pollinated by *Euglossa asarophora* Moure during September and October.

Distribution and habitat: Costa Rica. The present known distribution is limited to the lowland wet forests surrounding Gulfo Dulce, Costa Rica, the area near the Rincón airport on the Osa Peninsula, the forested slopes behind Golfito, and around Puerto Cortéz. This area harbors several other species of *Calathea* that apparently are endemic, such as *Calathea lasiophylla* and *C. longiflora.* The vegetation in this area is tropical wet (evergreen) forest which Allen (1956) called the Rain Forest of Golfo Dulce. *Calathea vinosa* typically occurs within the forest often on moderate to steep slopes.

Variation: The Osa Peninsula and Golfito populations are rather uniform in appearance. Cultivated specimens in more or less open garden situations are generally shorter than individuals growing in the forest. A plant quite similar to *Calathea vinosa,* but with a thinner, subglabrous leaf, was found in the greenhouse at Fairchild Tropical Gardens. Whether this is a variant of *C. vinosa* or represents a very closely related species is not known.

Relationships: *Calathea vinosa* is most closely related to *C. lasiophylla* and less so to *C. indecora.* The close resemblance of the inflorescences of *C. vinosa* and *C. lasiophylla* is striking, especially since the leaves are quite distinctive. The shape, pubescence and recurved margins of the bracts, and color and morphology of the flowers are similar. The habit, recurved bracts, and general form of the inflorescence of *C. vinosa* are similar to those of the more distantly related *C. silvicola.* The bracts of *C. silvicola* are usually subglabrous (puberulent in the Bocas del Toro populations), and the leaf is pleated rather than scalloped as in *C. vinosa.*

Common name: In Golfito, Costa Rica, *Calathea vinosa* is called café African.

Notes: The seeds germinate during the latter part of the rainy season, the new plants passing the dry season as seedlings.

7. **Calathea lasiophylla** H. Kennedy, sp. nov.

(Plate 11, Figure 14)

Planta robusta alta. Folia longe petiolata, petioli pars superior callosa teres glabra super puberula, ovata obtusa cum acumine base rotundata vel truncata et breviter acuminata supra nitida smaragdina glabra medianum puberula subtus opaca pallide viridi-grisea crebre pilosa; vagina subcoriacea, viridis sparse strigosa. Spica imbricata ovoido-cylindrica folio comitata, pedunculo viridi puberulo; bracteae spiraliter dispositae coriaceae, reniformae rotundatae saepe pallide virides velutinae recurvatae ceterum stramineae extus velutinae; paria florum ad 13 bracteolis membranaceis supra pilosis prophyllis et mesophyllis testaceis pilosis; ovarium eburneum glabrum praeter apicem sparse villosum; sepala eburnea apice pallide luteo; corollae tubus albus supra eburnea, lobi eburnei ovales acuti vel obtusi; staminodium exterius variabile album vel album pictum violaceum obovatum vel subspatulatum rotundatum vel bilobatum; callosum rectangulare obtusum vel praemorsum cum acumine cremeum apice purpurea maculata; cucullatum cremeum. Capsula turbinata obovata glabra straminea rotundata schistacea, arillo albo munita.

Caulescent herb 1–1.9 m. high. Shoots bearing 2–5 leaves, 0–2 basal, 2–4 cauline. Roots bearing root-tubers, these more prevalent in younger, nonflowering individuals. *Cataphylls* fleshy toward base, distal half coriaceous to chartaceous, ovate to narrowly ovate, acute, mucronulate, green, minutely tomentose, innermost 40–63 cm. long. *Leaf blade* semicoriaceous, ovate, apex obtuse with acuminate to protracted acuminate point, base rounded to truncate, junction with the pulvinus shortly acuminate, subtending leaf 38.3–53.8 cm. long and 17–27.6 cm. wide; others 53–78 cm. long and 22.5–33.6 cm. wide. Leaf surface above shiny, bright green, glabrous except for puberulent midrib; below, dull, light grey green densely pilose. *Pulvinus* olive green, glabrous except for line of hairs above, 3.1–11 cm. long, junction with petiole articulate, petiole lightly pigmented at junction, marking the junction with a light yellow green ring, extension of pulvinus along upper surface of petiole attenuate-acuminate. *Petiole* green with faint yellow green groove along upper side extending to apex of leaf sheath, sparsely strigose, in

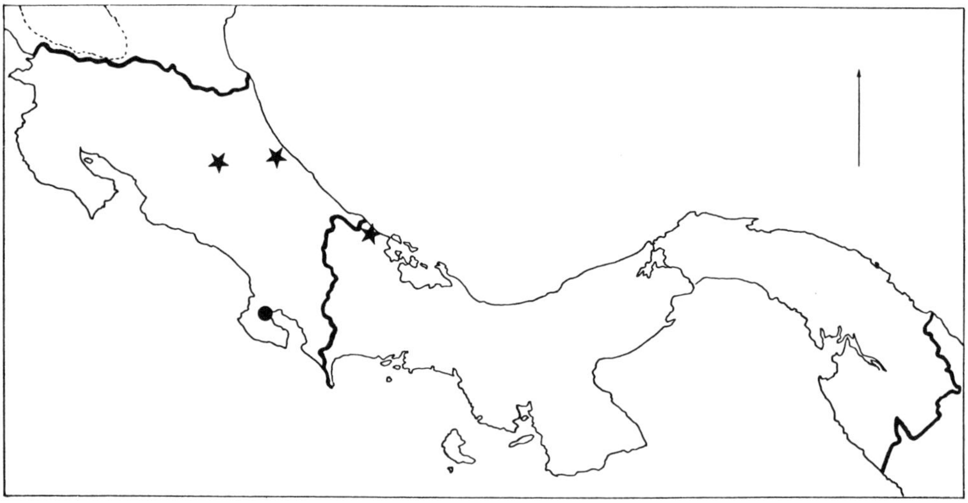

Fig. 14. Distribution of *Calathea foliosa* (stars) and *C. lasiophylla* (circles) in Costa Rica and Panamá.

subtending leaf, 0.9-6.7 cm. long; others 23-64 cm. long. *Leaf sheath* subcoriaceous, green, sparsely strigose, more densely so on the median abaxial surface and marginal edges, 12.5-21.8 cm. long in subtending leaf; others 38-82.8 cm. long. The petiole and leaf sheath of the subtending leaf usually puberulent rather than strigose. *Stem* green, subglabrous to sparingly puberulent, uppermost internode 20-34(60.7) cm. long, 1-3 elongated internodes present below the inflorescence. Nodal region usually lighter colored than adjacent stem portion, puberulent.

Inflorescence terminal, subtended by a foliage leaf, imbricate, ovoid-cylindrical, 5.5-13.5 cm. high and 5-6 cm. in diameter. *Peduncle* green, puberulent, (1.5)4.3-16 cm. long. *Bracts* 9-32, spirally arranged, slightly spreading at apex, apical margin reflexed to revolute, depressed-ovate to subreniform, apex rounded (occasionally first bract depressed ovate-triangular with acute apex), becoming very broadly ovate with obtuse apex in upper bracts, 3.3-4.6 cm. high and 5.1-7.2 cm. wide, each subtending up to 13 flower pairs. The first bract is much wider than following ones, the base attached almost completely around the peduncle. Outer surface pale straw colored at base, apical 1/2 pale greenish, velvety appressed pilose; inner surface shiny, cream, glabrous in lower 1/2, apical portion and especially margin pale greenish, densely velvety pilose. *Bicarinate prophyll* membranaceous, broadly obovate to transversely elliptic, translucent pale tan, appressed pilose on outer surface, glabrous within except for pubescent apex, 2.3-2.7 cm. high, height decreasing in apical bracts, 2.4-3.7 cm. wide, the first 1 or 2 decidedly wider than the rest. Tricarinate prophylls appressed pilose. *Secondary bracts* membranaceous, broadly obovate to transverse broadly obovate, apex truncate to subrounded, pale translucent yellow tan, basal 1/2 glabrous, appressed pilose above, 2.3-2.5 cm. high and 2.1-3.1 cm. broad. *Bracteoles* subtending individual flowers hyaline, pilose at apex, *ca.* 2-2.2 cm. high and *ca.* 3-5 mm. wide.

Flower bud never opens spontaneously. *Sepals* channeled, narrowly obovate, obtuse, basal half cream white, upper half faintly yellowish, sparsely pilose, (2.7)2.9-3.3 cm. long and 1-1.25 cm. wide. *Corolla tube* white below, cream above, glabrous, 2.85-3.2 cm. long; corolla lobes elliptical, acute to obtuse, cream white, 1.4-1.55 cm. long and 0.9-1 cm. wide. *Outer staminode* quite variable, obovate to subspatulate, irregular, rounded to emarginate, white with occasional streaks of purple, 1-1.2 cm. long and 0.55-0.75 cm. wide. *Callose staminode* rectangular, apex broadly obtuse to truncate with a point off center, cream color with a red to red purple patch at the center of the apex, 1-1.5 cm. long. *Cucullate staminode* cream colored, 1-1.2 cm. long and 4.5 mm. wide, provided with a small curved, channeled appendage subterminally, 1.5 mm. high. *Filament* with lateral petaloid appendage to 1.5 mm. wide; anther 3.5 mm. long. Style and stigma golden. *Ovary* cream white, glabrous with few scattered hairs at the apex, *ca.* 3 mm. high. *Capsule* thin, broadly obovate, rounded at the apex, smooth, straw colored, glabrous, *ca.* 1.3 cm. high and 0.9–1 cm. in diameter, crowned by a persistent live calyx, 3.4–3.7 cm. long and *ca.* 1.5–1.7 cm. wide. *Seeds* usually 3 per capsule, trigonous, rugose on outer surface, slate grey, 8–8.5 mm. high and 5.5 mm. in diameter, bearing a basal white aril 3 mm. high.

Type: COSTA RICA. Puntarenas: Osa Peninsula, 2.5 mi. S.W. of Rincón, forest across the Río Agua Buena from T.S.C. field station, 15–30 m., 8° 42′ N. x 83° 29′ W., 10 Sept. 1972, *H. Kennedy and F. Seifert 1600* (holotype F!; isotypes CR!, GH!, K!, MO!, US!).

Specimens seen: COSTA RICA. Puntarenas: Osa Peninsula, W. of Rincón de Osa, *ca.* mile 4 on the Pacific road, 15–30 m., 19 Oct. 1970, *H. Kennedy 648* (DUKE); 4 Sept. 1971, *H. Kennedy 1138* (COL, DAV, DUKE); rich primary forest with many palms near mile 11, primary forest and forest margins along Camino al Pacífico, W. of Rincón de Osa, 100 ft., 7 Aug. 1967, *P. Raven 21598* (F).

Flowering: *Calathea lasiophylla* flowers in the rainy season, July through November.

Distribution and habitat: Costa Rica. *Calathea lasiophylla* is known only from the type locality on the Osa Peninsula near the airfield southwest of Rincón and a few kilometers long the Pacific road. The vegetation of the area is tropical wet (evergreen) forest. This species is fairly shade tolerant, occurring within the forest, often near semiswampy areas with standing water but never actually in standing water or in decidedly waterlogged soil. Individuals of *C. lasiophylla* occasionally are found on steep stream banks.

Variation: One collection of *Calathea lasiophylla* from the Osa Peninsula near the type locality, *Jiménez 3031* (F 1672612), appears to belong to this species but differed in its puberulent major nerves on the upper leaf surface. The leaf is velvety tomentose below and the bracts densely pubescent as in the other collections. More material is desirable to ascertain the affinity of this population.

Relationship: *Calathea lasiophylla* is most closely related to *C. vinosa* and more distantly to *C. indecora.* All three species share a similar inflorescence and bract morphology. The inflorescence and flowers of *C. lasiophylla* resemble those of *C. vinosa,* and the bracts in both species are densely velvety tomentose with recurved to revolute margins. *Calathea indecora* has shorter bracts with margins that are somewhat recurved but never are revolute; the pubescence and shape of the bracts are similar. *Calathea lasiophylla, C. vinosa,* and *C. indecora* bear relatively large or long-petioled leaves which substantially exceed the height of the inflorescence.

Notes: The seeds germinate during the latter part of the rainy season, and the new plants pass the dry season as seedlings.

8. *Calathea indecora* Woodson

(Plate 12, Figure 13)

Calathea indecora Woodson, Ann. Missouri Bot. Gard. 29: 333. 1942. Type: Panamá, Prov. Bocas del Toro: Isla Colón, 15 Aug. 1940, *H. von Wedel 476* (MO 1239978!; isotype GH!).

Caulescent herb 1.22–3.0 m. high. Each shoot bearing (2)3–6 leaves, 1–4 basal and 1–3 cauline. *Cataphylls* usually 3, narrowly ovate, acute, mucronulate at apex, lower ones purple, upper ones dark green streaked with purple, glabrous to subglabrous, basally pilose, innermost 49–95 cm. high. *Leaf blade* narrowly ovate to narrowly elliptic, obtuse with acumen to acuminate in smaller leaves, base obtuse to rounded, abruptly short acuminate at junction with pulvinus, chartaceous, 23–52(82) cm. long and 7–23(31) cm. wide in subtending leaf, leaves 2 and 3, 37–89 cm. long and (9)14.3–37 cm. wide; when the plant bears as many as 5–6 leaves the lowest ones are distinctly smaller; 6.7–33 cm. long and 6.3–16 cm. wide. Leaf blade above dull green, glabrous, midrib yellow green with row of hairs apically, basal 1/2 subglabrous to glabrous; lower surface semishiny, greyish green, glabrous to tomentose along veins, more felt than seen, midrib yellow green, basal 1/3 tinged purple, tomentose. *Pulvinus* glabrous except for row of hairs along upper surface, olive green, light yellow green at junction with the petiole, may be tinged purplish below, 2.3–6(13) cm. long in subtending leaf, others (1.5)4–13.3 cm. long, junction with petiole articulate, acuminately extended along the petiole above. *Petiole* dull green to olive drab, glabrous to subglabrous, 6.3–28(87.3) in subtending leaf, others (1.1)26–102 cm. long. *Leaf sheath* chartaceous at margins, dull green, in lower leaves slightly to markedly streaked with purple, the junction with the stem often completely purple, margins and apex puberulent centrally glabrous, junction with stem sericeous, 5.5–17(33) cm. long in subtending leaves; others (15)17–108 cm. long. *Stem* internodes olive green, lower ones often streaked purple, glabrous.

Inflorescence terminal 97–164 cm. above ground level, subtended by a foliage leaf, the subtending leaf and 1 or 2 successive leaves held well above the inflorescence. *Peduncle* dull green minutely puberulent at and to 1.5 cm. below the junction with the inflorescence, the rest glabrous, 7–31 cm. long. Inflorescence ovoid to subcapitate 4.8–13 cm. high and 3.8–12.1 cm. wide, imbricate becoming somewhat lax with age. *Bracts* 12–36, coriaceous, reniform to depressed-ovate, apex rounded or rounded with an acumen or obtuse point centrally, the first bract usually bearing an acumen, basal portion imbricate, margin recurved to patent, 1.6–2.6(2.7) cm. high, the first bract 4.5–5.5 cm. wide, others 3.1–4.7 cm. wide, each subtending up to 10 flower pairs. Bract abaxially, where covered by next lower bract, cream, the upper portion including recurved adaxial margin pale greenish yellow, in younger inflorescences tinged light green, velvety tomentose throughout except adaxial imbricate portion, shiny glabrous. *Bicarinate prophyll* exceeding bract, transverse broadly obovate to obovate-elliptic, basal portion transparent, upper 1/2 light greenish yellow to pale yellow, sides and carina densely pilose, median area between glabrous to pilose, inner surface glabrous except apical 2–3 mm., 1.9–2.4 cm. high and 1.45–1.8 cm. wide, carina to carina, 2.8–3.5 cm. total width. *Secondary bracts* membranaceous broadly obovate, apex truncate, basal half transparent except for whitish veins, upper 1/2–1/3 translucent light yellow to light greenish yellow, pilose, outermost ones 2–2.2 cm. high and 2.2–2.6 cm. wide. *Bracteoles* subtending individual flowers narrowly obovate, channeled to carinate, apex pale yellow, glabrous, 1.6–2.4 cm. high, 0.4–0.7 cm. wide.

Flowers never opening spontaneously. *Sepals* channeled, semispongy at base and medially, narrowly obovate to oblong, apex obtuse, margins slightly infolded appearing acute, shiny, glabrous, basal 1/3 cream white, yellow above tinged faintly with green toward apex, 3.1–3.3 cm. long and 1–1.35 cm. wide. *Corolla tube* white, cream apically, glabrous, 3.3–3.4 cm. long, nectar holding area to 0.9 cm. long; *corolla lobes* unequal, narrowly subobovate to oblong, apex unequal in lower 2, equal in median upper lobe, obtuse to acute, light yellow at base, greenish yellow toward apex, glabrous, median petal narrower than lateral ones; 1.3–1.6 cm. long and 0.55–0.8 cm. wide. *Outer staminode* obovate to unguiculate, slightly cupped apically, rounded to emarginate, 1.2–1.3 cm. long and 0.7–0.8 cm. wide, occasionally absent. *Callose staminode* rectangular, apex bidentate to one side, light golden orange. *Cucullate staminode* golden yellow orange basally, hood portion cream, 0.9–1 cm. long and 4.5–5.5 mm. wide, bearing a channeled fingerlike appendage subterminally to 1.5 mm. high. *Filament* with narrow lateral appendage extending to tip of anther, portion attached to anther narrower than filament portion; anther 3 mm. long. Style and stigma deep golden. *Ovary* cream, glabrous, 2–4 mm. long. *Capsule* obovoid, thin, straw colored, smooth, glabrous, 1.4–1.6 cm. high and 0.75–0.85 cm. wide, crowned by a persistent live calyx 1.3–1.6 cm. wide at base. *Seeds* usually 3 per capsule, trigonous, outer surface rugose, slate grey, 7–8.5 mm. high and 4–5 mm. wide, bearing a basal white aril 3–4 mm. high.

Representative specimens seen: PANAMÁ. Bocas del Toro: Bastimentos, northern side of island, terra firma to slightly swampy, 8 Sept. 1969, *H. Kennedy and D. Dreyfus 369* (DAV); filo de Almirante W.S.W. of Almirante, 19 Nov. 1971, *H. Kennedy and R. L. Dressler 1290* (DUKE, F, K, MO, NY, PMA, US); Water Valley, 11 Sept. 1940, *H. von Wedel 712* (BM); Isla Colón, 18 Oct. 1940, *H. von Wedel 1229* (MO, GH) both the GH specimen and the MO one have the same collection number but the label for the MO specimen reads Old Bank Isl.; Old Bank Isl., 15 Feb. 1941, *H. von Wedel 2101* (MO); Bastimentos Isl., 5 Nov. 1941, *H. von Wedel 2908* (MO); without exact locality, 29 July 1940, *H. von Wedel 209* (BM). Prov. unknown: 1927, *C. E. Moore 10* (C, F, S).

COSTA RICA. Limón: in *Gynaerium sagittatum* thickets along Río Reventazón below farmhouse, Finca Castilla, 30 m., 27 July 1936, *C. W. Dodge and V. F. Goerger 9413* (CR, F, GH, MICH, MO, US); Hamburg Finca, on the Río Reventazón below Cairo, *ca.* 55 m., 19 Feb. 1926, *P. C. Standley and J. Valerio 48860* (US); *P. C. Standley and J. Valerio 48865* (US). San José: steep wet slopes of old secondary growth and forested streamsides along Río Claro Valley at 1000–1200 m., (Río La Hondura drainage) below La Palma, N.E. of San Jerónimo, 10° 3′ N. x 83° 58′ W., 12 Sept. 1971, *W. C. Burger and M. Burger 8498* (F); below Bajo La Hondura, Río La Hondura drainage, steep slopes, trail below the finca, 1100 m., 6 Sept. 1972, *H. Kennedy 1563* (CR, F, GH, MO, US). Alajuela: low tropical rain forest, with palms prominent in understory, vicin. of Guatuso de San Rafael (on Río Frio), 10° 43′ N. x 84° 48′ W., 80–100 m., 4 Aug. 1949, *R. W. Holm and H. H. Iltis 870* (A, P, MO).

Flowering: Calathea indecora flowers mainly during the late rainy season, October through December, although flowering individuals have been collected from July through February.

Pollinator: Populations of *C. indecora* near Almirante in Bocas del Toro, Panamá, were observed to be pollinated by *Euglossa ignita* Smith in November, 1971.

Distribution and habitat: *Calathea indecora* occurs predominantly in the Atlantic lowlands. Its southernmost locality is in Bocas del Toro, Panamá, but it ranges northward along the coast into Limón Province, Costa Rica, in tropical wet (evergreen) to premontane wet forest. The highest elevation at which the species was collected is *ca.* 1100 m. near Bajo La Hondura, Costa Rica, in premontane rain forest; the northern end of the range is near Guatuso de San Rafael, Alajuela Province, Costa Rica. The lowland populations often occur in large stands in poorly drained semiswampy areas but are never actually in standing water. They often grow in full sun. In the forest behind Almirante, Panamá, and in the Bajo La Hondura area in Costa Rica, the plants are scattered within the forest on moderate slopes.

Variation: The lowland populations of *Calathea indecora* appear rather uniform in qualitative characters. Plants from Bajo La Hondura, Costa Rica, tend to have fewer leaves overall, usually two to four as opposed to four to five for the Bocas del Toro, Panamá, population. The veins on the lower leaf surface of the Bajo La Hondura population are subglabrous to glabrous, whereas those of the Bocas del Toro population are definitely puberulent. The bracts of the Bocas plants are usually rounded, and the apical margin is patent to slightly recurved in upper bracts. The plants from Bajo La Hondura and the San Rafael area commonly have a short obtuse apical extension of the recurved bract. The margin of the bract in general tends to be more recurved in the last two populations.

Relationship: *Calathea indecora* is most closely related to *C. vinosa* and *C. lasiophylla*, sharing a similar inflorescence and flower morphology, and pubescence and texture of bracts. It is more distantly related to *C. foliosa* which has similar but more strongly recurved bracts than *C. indecora* and has no basal leaves.

Notes: The seeds germinate during the latter part of the rainy season and the young plants pass the dry season as seedlings.

9. **Calathea silvicola** H. Kennedy, sp. nov.

Planta caulina robusta elata. Folia longe petiolata, petioli pars superior callosa teres glabra vel minute tomentosa, plicata ovata acuminata vel breviter acuminata base rotundata vel truncata supra viridia glabra praeter apicem interdum subciliatum nervi tomentosi medianum puberula frequenter ruber subtus pallide viridia vel atropurpurea venae et medianum tomentosa ceterum glabra vel ubique tomentosa herbacea; vagina herbacea vel chartacea minutissime puberula viridis vel purpurea. Spica ovoidea vel subcylindrica pedunculo subtomentoso sustenta, bracteae spiraliter dispositae coriaceae depresse ovatae acutae vel rotundatae apice purpureo vel viridis recurvato ceterum virides vel luteae extus minutissime pilosae vel tomentosae; flores cleistopetali paria florum 6 vel ultra bracteolis lanceolatis hyalinis membranaceis mesophyllisque comitata; ovarium album glabrum; sepala anguste elliptica vel anguste obovata acuta vel rotundata; corollae tubus albidus; lobi albidi vel pallide luteo-aurantiaci ovati vel elliptici acuti vel obtusi; staminodium exterius album obtriangularum bi- vel tridentatum, cucullatum cremeum. Semina rugosa trigona dorso rotundata, schistacea, arillo albo munita.

This species occurs primarily in the wet (evergreen) forest zone from the Department of El Valle in Colombia north to Costa Rica. The plants in Bocas del Toro, Panamá, and Limón Province, Costa Rica, are recognized as a distinct geographic subspecies.

Key to the subspecies of *Calathea silvicola*

a. Pulvinus glabrous except for row of hairs above; leaves 1-3; sepals and bracts green occasionally tinged dark purple . . . . subspecies *silvicola*

aa. Pulvinus minutely puberulent (14X) throughout; leaves 3-7; sepals and bracts yellow . . . . subspecies *tomentosa*

9a. **Calathea silvicola** ssp. **silvicola**

(Plate 13, Figure 15)

Differt ab aliis speciebus floribus cleistopetalis subgeneris *Pseudophrynium* ser. *Scapifoliae* foliis 1-3 plicatis, bracteis viridibus marginibus patentibus vel recurvatis instructis, his demum purpurascentibus et petiolis purpureis aut viridibus atque purpureo-lineatis et caulibus purpureis.

Caulescent herb 1-1.8 m. tall, bearing 1-3 leaves; 1-2 cauline, 0-2 basal. Rhizome 1.5-2.3 cm. in diameter, internodes 5-11 mm. *Cataphylls* usually 2, dark purple, tomentose 8-30 cm. long and 2.5 cm.

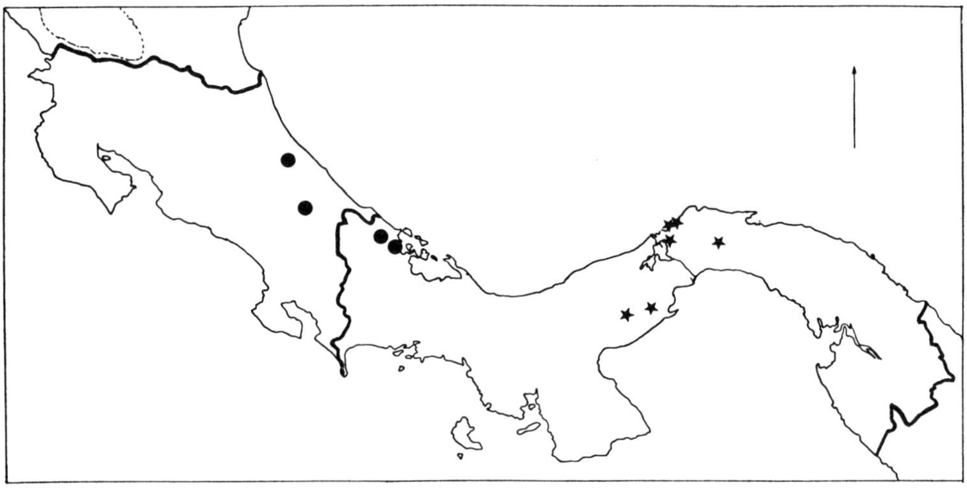

Fig. 15. Distribution of *Calathea silvicola* ssp. *silvicola* (stars) and *C. silvicola* ssp. *tomentosa* (circles) in Costa Rica and Panamá.

wide. *Leaf blade* herbaceous, appearing shallowly pleated, ovate, apex acuminate to abruptly acuminate, base rounded to truncate, 39-62 cm. long and 19-32.5 cm. wide. Uppermost leaf subtending the inflorescence somewhat smaller than the lower leaves. Upper surface slightly shiny, grass green frequently with red midrib in basal half, glabrous except pubescent midrib, more densely so toward apex; the lower surface light green or dark purple, midrib pale green, red, or dark purple, tomentose along major veins (or both major and minor) and sides of the midrib, or minutely puberulent throughout. *Pulvinus* slightly oval in cross section, light green to purple, or green striped with purple, glabrous, 2.8-5.1 cm. long. *Petiole* purple or green finely striped with purple, minutely puberulent, (1.5)11.5-53 cm. long. *Leaf sheath* herbaceous, purple, occasionally green, tomentose, 9-50 cm. long, sheath of the leaf subtending the inflorescence often quite short, not auriculate at the apex. *Stem* purple to purple green, tomentose.

Inflorescence terminal, subtended by a foliage leaf, imbricate, ovoid, 6-9 cm. high and 4-5.5 cm. in diameter. *Peduncle* green, tomentose, especially at the junction with the inflorescence, 5.5-23 cm. long. *Bracts* 12-30 (occasionally more), spirally arranged, slightly spreading, reflexed at the apex, transversely ovate, apex acute to obtuse, 2-3.2 cm. high and 3.5-4.5 cm. wide, the uppermost bracts narrower and longer than lower ones, each subtending 6 or more flower pairs. Abaxial surface of bract green, sparingly pilose; adaxial surface green, glabrous below, reflexed margin dark green or tinged purple, occasionally plain green, sparingly pilose. *Bicarinate prophyll* membranaceous, broadly elliptic to transverse broadly elliptic, apex rounded, sparsely puberulent at apex, *ca.* 2.3-2.6 cm. high and *ca.* 2.3-2.4 cm. wide. *Secondary bracts* membranaceous, transverse broadly obovate-elliptic, apex truncate to slightly rounded occasionally shallowly retuse, apical margin minutely puberulent, *ca.* 1.9-2 cm. high and 2.2-2.3 cm. wide. *Bracteoles* subtending individual flowers *ca.* 2.1-2.25 cm. high and *ca.* 0.35-0.45 cm. wide.

Flower never opens spontaneously. *Sepals* semispongy at base, channeled, narrowly obovate, obtuse to acute, green above, white at base, glabrous, 2.6-3 cm. long and 0.7-0.8 cm. wide. *Corolla tube* white, glabrous, 2.7-2.9 cm. long; corolla lobes subequal, white, ovate to elliptic, obtuse, 1.4-1.6 cm. long and 0.5-0.8 cm. wide. *Outer staminode* narrowly obtriangular, apex tridentate, the central tooth smaller, white, 1-1.1 cm. long and 0.35-0.45 cm. wide. *Callose staminode* rectangular, white, 0.9-1 cm. long. *Cucullate staminode* white, 0.9-1.1 cm. long, provided with a fingerlike, channeled appendage subterminally. *Filament* of stamen with narrow lateral petaloid appendage, anther 3.5-4 mm. long. *Ovary* white, glabrous, 2.5 mm. high. Style and stigma golden. *Capsule* obovoid, cream to whitish, 1.05-1.35 cm. high and 0.8-0.9 cm. wide, crowned by a persistent calyx, sepals increasing to over 3 cm. long and 1.4 cm. wide, becoming tinged with purple with age. *Seeds* usually 3 per capsule, slate grey, trigonous, rugose on the outer surface, 6-8.5 mm. high and 4-5 mm. wide, provided with a basal white aril 2.5-4 mm. long.

Type: PANAMÁ. Colón: near Portobelo, by the bridge over Río Buenaventura, 14 Aug. 1970, *H. Kennedy and R. B. Foster 451A* (holotype US!; isotype F!).

Specimens seen: COLOMBIA. El Valle: Cordillera Occidental, vertiente occidental, Hoya del Río Anchicayá, lado derecho, bajando a la Planta, bosques, 200–350 m., 27 Sept. 1943, *J. Cuatrecasas 15208* (COL, F, US); Anchicayá, at the hydroelectric plant, 22 Jan. 1971, *H. Kennedy 751* (US). Chocó: bosque al lado derecho del Río Condoto, frente a Condoto, 70 m., 23 Aug. 1955, *J. M. Idrobo 1845* (COL); Río Taparal off Río San Juan, 100 ft., 25 Aug. 1962, *J. W. L. Robinson 345* (K).

PANAMÁ. Darién: camp summit, adjacent Darién-San Blas border along the sea level canal route 17, seasonal evergreen forest and weedy clearing, 1000–1200 ft., 18 Dec. 1967, *R. L. Oliver et al. 3666* (MO). Panamá: Cerro Campana, *ca.* 800 m., 18 Sept. 1968, *H. Kennedy, R. B. Foster, R. Weaver 78* (COL); La Eneida, region of Cerro Jefe, *ca.* 750 m., 8 July 1969, *H. Kennedy, R. L. Dressler, N. H. Williams 290* (F); region of Cerro Azul, Finca El Indio, *ca.* 750 m., 17 July 1969, *H. Kennedy 306* (BM, F); Cerro Campana, *ca.* 800 m., cloud forest, 1 Sept. 1970, *H. Kennedy 490* (F); La Eneida, region of Cerro Jefe, *ca.* 750 m., 13 Sept. 1970, *H. Kennedy 497* (PMA); region of Cerro Azul, Finca El Indio, *ca.* 750 m., 29 Aug. 1972, *H. Kennedy 1515* (DUKE, MICH, MO, UC); Cerro Campana, *ca.* 1 mi. past Florida State field station, *ca.* 800 m., 3 Sept. 1972, *H. Kennedy 1540* (F); Cerro Campana, cloud forest, 8 Aug. 1970, *J. L. Luteyn and H. Kennedy 1839* (DUKE). Colón: Santa Rita Ridge, 5 mi. from highway, 31 Jan. 1971, *T. Croat 13199* (MO); Santa Rita Ridge Lumber Road, 10 July 1969, *H. Kennedy, R. L. Dressler and N. H. Williams 294* (COL, DAV, K); Río Guanche *ca.* 1.5 mi. upstream from the bridge, *ca.* 10 m., 10 Aug. 1971, *H. Kennedy, R. L. Dressler, H. Wiehler 1101* (U); 23 Aug. 1972, *H. Kennedy and R. L. Dressler 1503* (PMA, BR); Santa Rita Ridge below the rain gauge, 13 Aug. 1971, *H. Kennedy, R. L. Dressler, H. Wiehler 1128* (PMA). Coclé: near La Mesa, El Valle de Antón, cloud forest, *ca.* 800 m., 2 Aug. 1970, *H. Kennedy and J. L. Luteyn 450* (GH, US).

Flowering: *Calathea silvicola* ssp. *silvicola* flowers during the rainy season, July through November. A specimen collected in January was past flowering.

Pollinator: *Calathea silvicola* ssp. *silvicola* was observed to be pollinated by *Eulaema speciosa* (Moscary) on Cerro Jefe and Cerro Campana, Panamá.

Distribution and habitat: *Calathea silvicola* ssp. *silvicola* is widespread and is known from the Dept. El Valle, Colombia, in the south to near El Valle, Coclé Province, Panamá, in the north. It occurs in tropical wet (evergreen) to premontane rain forests within the forest in moderate shade and often grows on a slight slope but may also be found on level areas or occasionally near stream sides.

Variation: The amount of variation in *Calathea silvicola* ssp. *silvicola* is rather small and is mostly in leaf pubescence and color. In the majority of plants, the leaf is light green below; however, two individuals from La Mesa near El Valle, Panamá, had a wholly dark red purple underleaf. The bract margins usually change from green to purple with age. Whether the apex of the leaf is rounded with a point or acuminate depends on the size of the leaf and varies within populations. Plants from Cerro Campana and El Valle, Panamá, have puberulent veins on the lower leaf surface, and the intervein tissue is glabrous, while those from Cerro Jefe and La Eneida, eastern Panamá, are minutely puberulent throughout. Both patterns of leaf pubescence are present in the population from Río Guanche, Panamá, while in the population from Anchicayá, Colombia, only the veins are puberulent. The upper surface of the leaf blade is minutely puberulent near the base in the population from the Río Guanche, Panamá, but it is glabrous in other populations. The upper leaf surface is commonly sparse tomentose in the juvenile plants.

Relationship: The habit, inflorescence morphology, and floral morphology of *Calathea silvicola* ssp. *silvicola* are similar to those of *C. vinosa* and *C. indecora*. Neither of the last two species show the color change to purple with age that is characteristic of *C. donnell-smithii* and its related species. Thus, these latter species are probably related to *C. silvicola*, though more distantly than are *C. vinosa* and *C. indecora*.

9b. *Calathea silvicola* ssp. **tomentosa** H. Kennedy, ssp. nov.

(Plate 14, Figure 15)

Morphologia foliorum *C. silvicolae* ssp. *silvicolae* simillima, habitu inter *C. silvicolam* ssp. *silvicolam* et *C. indecoram* intermedia; a *C. silvicola* ssp. *silvicola* et *C. indecora* differt venis puberulis in parte apicali superioris folioram et pulvino puberulo; a *C. silvicola* ssp. *silvicola* differt foliis tribus vel septem et bracteis pallide flavis marginibus recurvatis viriscente-flavis et caulibus viridibus.

Caulescent rhizomatous herb, 0.9–1.7 m. tall. Shoot bearing 3–7 leaves, 1–3 cauline, 1–4(6) basal. *Cataphylls* 2–3, chartaceous, thicker toward base, narrowly ovate, acute, mucronulate, purplish or purplish green, densely minute tomentose, innermost 33–62 cm. high. *Leaf blade* chartaceous, pleated, major veins raised, elliptic to ovate, apex obtuse with a point or abruptly acuminate, base obtuse to rounded, 29–43 cm. long and 14.5–25.7 cm. wide, in subtending leaf; others 26.5–57.5 cm. long and 16–30.5 cm. wide. Leaf above glittering grass green to deep green, subglabrous, veins near apex puberulent, apical margin nearly ciliate, midrib lighter green, puberulent centrally; lower surface light grey green, tomentose, midrib yellow green, minutely tomentose. *Pulvinus* yellow green to light olive green, minutely puberulent, felt more than seen, more dense above, acuminately extended along the upper surface to 3 cm., 3–5 cm. long in subtending leaf; others 5–8 cm. long. *Petiole* dull, deep green, tomentose to subscabrous, 8–17 cm. long in subtending leaf; 24–69 cm. long in others. Leaf sheath chartaceous, not auriculate, green, darker toward base, minutely tomentose, 13–18 cm. long in subtending leaf; 30–59 cm. long in others. *Stem* deep green, minutely tomentose.

Inflorescence terminal subtended by a foliage leaf, ovoid to subcylindric, imbricate, 7–9.3 cm. high and 4–6 cm. wide. *Peduncle* green, minutely puberulent 10–14 cm. long. *Bracts* 16–28, coriaceous, first bract or two reniform, the rest depressed-ovate, apex rounded to occasionally obtuse in upper bracts, margins patent to revolute, 2.3–3.4 cm. high and 4.5–6.5 cm. wide. Outer surface light yellow with tinge of yellow orange, tomentose, adaxial surface shiny light yellow, glabrous basally, apical recurved margin chartreuse, tomentose. *Bicarinate prophylls* membranaceous, rotund to transverse broadly elliptic, apex rounded, translucent yellow, apical margins and center sparsely minute puberulent, 1.9–2.1 cm. high and 2.1–3.1 cm. wide. *Secondary bracts* membranaceous, translucent yellow, broadly obovate to transversely broadly ovate, retuse, sparsely minute puberulent at apex, 1.8–2.1 cm. high and 1.7–2.2 cm. wide. *Bracteoles* membranaceous, translucent light yellow, 1.8–2.1 cm. long and 0.25–0.45 cm. wide.

Flowers never open spontaneously. *Sepals* coriaceous, base and central portion semispongy, narrowly obovate, obtuse to rounded, shiny cream white at base, yellow above, tinged faintly greenish at tip, glabrous, 2.7–3.1 cm. high and 1–1.2 cm. wide. *Corolla tube* white to cream above, glabrous, *ca.* 3.3–3.5 cm. long; corolla lobes unequal, elliptic to broadly elliptic, obtuse apically, light yellow orange, 1.4–1.5 cm. high and 0.9–1.15 cm. wide. *Outer staminode* narrowly obovate to obtriangular, apex bilobed, cream to white, *ca.* 1.1–1.2 cm. long and *ca.* 0.6–0.75 cm. wide. *Callose staminode* rectangular, golden orange *ca.* 1.1–1.2 cm. high. *Cucullate staminode* cream at tip, light golden orange basally, *ca.* 0.9–1.05 cm. long and *ca.* 0.5–0.55 cm. wide, bearing a fingerlike channeled appendage subterminally. *Filament* bearing a narrow lateral petaloid appendage to 1.5 mm. wide, upper 1.5 mm. of filament free; anther *ca.* 3.5 mm. long, appendage attached throughout its length. Style and stigma golden orange. *Ovary* white, glabrous, 3–3.5 mm. high. *Capsule* obovoid, pale cream yellow, crowned by a persistent, live calyx, mature size unknown. *Seeds* usually 3 per capsule, slate grey, provided with a basal white aril.

Type: PANAMÁ. Bocas del Toro: near Almirante, 22 May 1972, *R. L. Dressler 4191* (holotype US!).

Specimens seen: PANAMÁ. Bocas del Toro: premontane rain forest between Quebrada Lugron and Cerro Bonyic, near Río Terebe, 300–900 ft., 13 April 1968, *J. H. Kirkbride Jr. and J. A. Duke 629* (MO).

COSTA RICA. Limón: Río Hondo, plains of Santa Clara, 100 m., 7 May 1903, *O. F. Cook and C. B. Doyle 582* (US); Talamanca, forêts de Tsâki, 200 m., March 1895, *A. Tonduz 9437* (BR).

Flowering: *Calathea silvicola* ssp. *tomentosa* flowers in the early rainy season, March through July; the beginning of flowering may be delayed following an unusually dry or long dry season.

Distribution and habitat: *Calathea silvicola* ssp. *tomentosa* is known only from Bocas del Toro Province, Panamá, and Limón Province, Costa Rica, in the wet (evergreen) forest. Plants of this subspecies usually occur as clumps rather than as scattered, isolated individuals. In contrast to ssp. *silvicola,* ssp. *tomentosa* does not often occur within a mature forest but is found in secondary or disturbed situations. In Bocas del Toro ssp. *tomentosa* is most commonly in association with *Heliconia,* weedy *Piper* spp., and weedy Urticaceae.

Variation: There seems to be as much or more variation between plants of *Calathea silvicola* ssp. *tomentosa* within an area than between those of different localities. There is a noticeable difference in the density of pubescence on the inflorescence bracts; however, they are more densely pubescent than in ssp. *silvicola.* Often the entire inner surface of the bract is minutely puberulent, whereas in ssp. *silvicola* the lower half of the bract is glabrous. The underside of the leaf is minutely tomentose throughout, whereas in ssp. *silvicola* often only the veins are tomentose. The degree to which the bract margin is reflexed in Bocas del Toro plants ranges from a strongly reflexed revolute margin, as in *C. vinosa*, to subpatent. Subspecies *silvicola* is variable to a lesser degree; its bract margins are rarely revolute. Subspecies *tomentosa* differs in habit from ssp. *silvicola*: the former subspecies has several leaves, one of which is always basal, while ssp. *silvicola* often has only a single leaf (the subtending leaf) and frequently only has two leaves. The color of the bracts and flowers is one of the most striking differences between the two subspecies. The bracts, sepals, and petals are yellowish in ssp. *tomentosa* and the bracts are green, often with a purple margin; the sepals are green and the petals are white in ssp. *silvicola*. The leaf sheath is green in ssp. *tomentosa* and is usually purple in ssp. *silvicola*.

Relationship: The more tomentose bracts, more numerous leaves, pubescence of the bicarinate prophyll, and retuse apex of the secondary bracts suggest that *Calathea silvicola* ssp. *tomentosa* has a closer affinity to *C. indecora* than does ssp. *silvicola*. It is possible that ssp. *tomentosa* is the result of hybridization between *C. silvicola* ssp. *silvicola* and *C. indecora*. However, ssp. *tomentosa* has yellowish bracts and sepals, characters that neither *C. indecora* nor *C. silvicola* ssp. *silvicola* possesses.

Notes: Two flowering individuals observed near Almirante had an aberrant flower structure. The flowers had two outer staminodes, the first in the normal position between the cucullate staminode and the stamen and the second located between the stamen and the callose staminode. The second abnormal structure was a narrow, somewhat callose staminode, *ca.* 1 cm. long, that was located between the callose staminode and the cucullatum. Thus all of the staminodes in the outer whorl were represented.

10. *Calathea foliosa* Rowlee ex Woodson

(Plate 15, Figure 14)

Calathea foliosa Rowlee ex Woodson, Ann. Missouri Bot. Gard. 29: 332. 1942. Type: Farm 6, near Almirante, Prov. Bocas del Toro, Panamá, 23 Sept. 1920, *N. W. Blair 1016* (holotype US 1315514!).

Caulescent herb 1.6-2 m. high. Each shoot bearing 3-7(11) cauline leaves above an elongated stem internode 1-1.35 m. long. The leaves held in a spiral arrangement. *Cataphylls* usually 2, narrowly ovate, acute, mucronulate at apex, green tinged with purple giving a dark green to purple green color, glabrous, persistent live in flowering individuals, innermost 54-62 cm. long. *Leaf blade* ovate to narrowly ovate or subelliptic, acuminate, base obtuse to rounded, 16-44 cm. long and 6.5-16.1 cm. wide. Upper leaf surface bright grass green, shiny, especially so in young leaves, glabrous except for puberulent midrib; lower surface dull, light grey green, glabrous. *Pulvinus* olive green, glabrous except for row of hairs along upper surface, 0.8-2.3 cm. long. *Petiole* green, usually absent, occasionally up to 1.3 cm. long. *Leaf sheath* green streaked to brushed purple, especially toward the base, dark purple green at junction to stem, margins chartaceous, glabrous except outer margins puberulent, especially toward base, 11-20 cm. long in subtending leaves; others 16.5-45 cm. long. *Stem* dark green, glabrous. Stem internodes between the leaves have a slightly zigzag appearance, the upper portion of the leaf sheath becoming free from the stem just at the point where the next higher leaf arises as opposed to clasping the stem throughout its length.

Inflorescence terminal, subtended by a foliage leaf, imbricate, capitate, subcylindrical to globose, 4.8-7.5 cm. high and 4.5-6.8 cm. wide. *Peduncle* pale green, tomentose to subglabrous, 2.3-5(7) cm. long. *Bracts* 9-24, coriaceous, first bract reniform, apex rounded with acumen or abruptly caudate-acuminate centrally, second and lower bracts broadly ovate to broadly trullate (1:1 ratio), apex acuminate, upper bracts becoming trullate with an acute apex, basal 2/3 clasping, apical margin recurved, 2.5-4.8 cm. high and 3.1-4.6 cm. wide. Abaxial clasping portion straw yellow, recurved margin lighter, adaxial clasping portion shiny pale yellow; recurved margin light yellow with slight tinge of green, with age margin frequently turning dark purple and the basal abaxial portion tinged purplish, velvety tomentose throughout except clasping adaxial portion, glabrous. Each bract subtends at least 6 flower pairs. *Bicarinate prophyll* membranaceous, elliptic to transverse broadly elliptic, apex rounded, translucent straw color, with age apex and back turning purple, pilose, *ca.* 2-2.2 cm. high and *ca.* 1.8-2.3 cm. wide. *Secondary bracts* membranaceous, broadly obovate-oblong, apex rounded, translucent straw color, with age apex tinged purple, pilose above, subglabrous below, *ca.* 2-2.2 cm. high and *ca.* 2-2.2 cm. wide. *Bracteoles* subtending individual flowers membranaceous, pilose apically, *ca.* 1.6-1.8 cm. high and *ca.* 2.5-3 mm. wide.

Flowers never opening spontaneously. *Sepals* semispongy in basal and central portion, narrowly obovate, apex obtuse to rounded, pale yellow at base to bright yellow above, glabrous, 3.2-3.4 cm. high and

0.75–0.9 cm. wide. *Corolla tube* cream, glabrous, 3.9–4.2 cm. long; corolla lobes unequal, elliptic to broadly elliptic, apex obtuse, margins inrolled appearing acute, light golden yellow, 1.6–1.9 cm. long and 0.65–0.9 cm. wide. *Outer staminode* cuneiform to narrowly obtriangular, emarginate, pale cream white, 1.2–1.4 cm. long and 0.7–1.1 cm. wide. *Callose staminode* rectangular, apex bidentate to somewhat irregular, golden, 1.1–1.3 cm. high. *Cucullate staminode* cream, basal portion golden, *ca.* 0.95 cm. long and *ca.* 0.5 cm. wide, bearing a fingerlike channeled appendage subterminally. *Filament* bearing narrow lateral appendage 1 mm. wide, upper 2 mm. free from style; anther 3.5–4 mm. long, upper 3/4 free. Style and stigma golden. *Ovary* cream, glabrous 2–2.5 mm. high. *Capsule* trigonous, obovoid, cream white with yellow tinge above, 1.1–1.3 cm. high and 0.9–1.05 cm. wide, crowned by a persistent live calyx; sepals tinged purple at the apex with age, on mature capsules dark purple, expanded to 0.95–1.1 cm. wide at the base. *Seeds* usually 3, trigonous, outer surface rugose, slate grey, 7–8 mm. high and 5.5–6 mm. in diameter, bearing a white aril *ca.* 3 mm. long.

Specimens seen: COSTA RICA. Limón: forêts de Río Hondo, Santa Clara, 100 m., Sept. 1903, *H. Pittier 16727* (US). San José: steep wet slopes and forested stream sides, Río Claro Valley at 1000–1200 m., (Río La Hondura drainage) below La Palma, N.E. of San Jerónimo, 10° 3′ N. x 83° 58′ W., 19 Nov. 1969, *W. C. Burger and R. Liesner 6263* (CR, F); steep wet slopes of old secondary growth and forested streamsides along the Río Claro Valley at 1000–1200 m. (Río La Hondura drainage) below La Palma, N.E. of San Jeronimo, 10° 3′ N. x 83° 58′ W., 12 Sept. 1971, *W. C. Burger and M. Burger 8495* (F); below Bajo La Hondura, Río La Hondura drainage, steep slopes, trail below the finca, 1100 m., 6 Sept. 1972, *H. Kennedy 1559* (CR, MO, US); trail beyond Bajo La Hondura towards Río Claro, cloud forest at approx. 1100–1200 m., 4 July 1972, *J. L. Luteyn 3322* (DUKE).

Flowering: *Calathea foliosa* flowers during the rainy season, from July through November.

Distribution and habitat: The holotype of *Calathea foliosa* was collected from the southern- and easternmost locality known for the species, near Changuinola, Panamá. Another lowland collection was made in Limón Province, Costa Rica, near Río Hondo. The vegetation in both of these lowland areas is tropical evergreen forest (Holdridge's [1967] premontane moist to premontane wet forest). The northernmost collections are from a mid-elevation (*ca.* 1100 m.) premontane rain forest on the slopes of the Río La Hondura drainage, San José Province, Costa Rica. *Calathea foliosa* is found within the forest and usually grows along the upper banks of stream channels. The type locality is presently planted in bananas and collecting in the surrounding areas failed to uncover additional plants of this species.

Variation: Little variation occurs in the few known collections of *Calathea foliosa.* There is a striking difference, however, between the type specimen and those from near Bajo La Hondura, Costa Rica, in the arrangement of leaves. The leaves of the type are clustered at the base of the peduncle, and the internodes between successive leaves are 3–6 mm. long, whereas the internodes of the other specimens are several centimeters long. The basal margins of the leaf sheath at its junction with the stem are reflexed away from the stem rather than clasping and are thickened and rough at that point. Despite the obvious abnormality of the habit of the type, it seems clear that it and the Costa Rican plants constitute a single species. The characteristic features in common are the densely pubescent bracts with recurved margins, pubescent bicarinate prophylls, acuminate leaf tips, frequent absence of petiole, and the strictly cauline position of the leaves. Due to the condition of the type, the leaf shape is not apparent.

Relationship: The general habit, the leaf morphology, and the color change with age in the bracts and sepals of *Calathea foliosa* suggest that it is most closely related to *C. venusta* and *C. donnell-smithii.* It is more distantly related to *C. indecora,* sharing a similar bract texture and pubescence with that species. The individuals of *C. indecora* near Bajo La Hondura, Costa Rica also have a reflexed bract margin.

Notes: The leaf sheaths of the holotype of *Calathea foliosa* have hardened areas (appearing somewhat swollen in dried condition) that frequently have small holes in them. The same type of damage is seen in specimens from Bajo La Hondura, Costa Rica. These appear to be the result of insect attack. Such injury to the plant may have been the cause of the abnormal leaf arrangement which characterizes the holotype.

11. *Calathea venusta* H. Kennedy

(Plate 16, Figure 16)

Calathea venusta H. Kennedy, Ann. Missouri Bot. Gard. 60: 413–426. 1973. Type: Río Guanche, mature forest about 1.5 mi. upstream from bridge, *ca.* 10 m., Prov. Colón, Panamá, 1 Nov. 1971, *H. Kennedy 1234* (MO!; isotypes F!, US!).

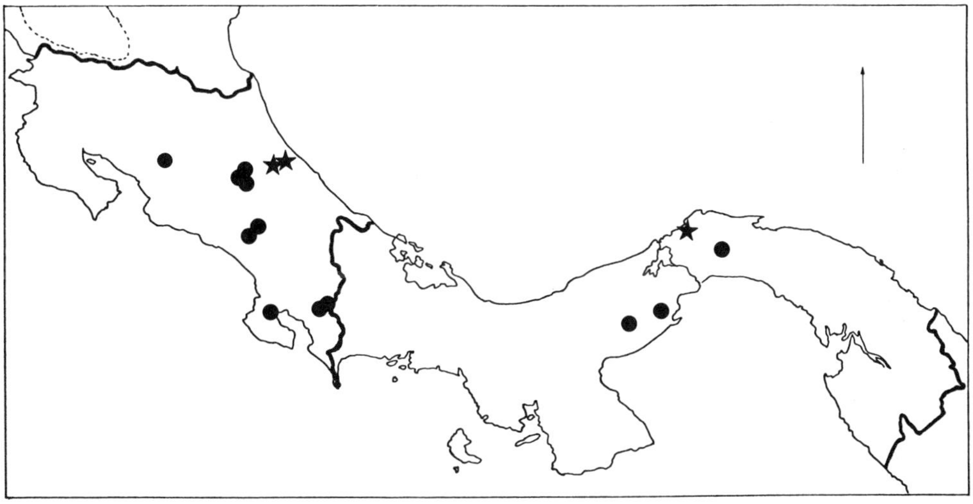

Fig. 16. Distribution of *Calathea venusta* (stars) and *C. donnell-smithii* (circles) in Costa Rica and Panamá.

Cauline herb 1.14–1.43 m. high. Juveniles and subadults bear 9-17 leaves; the uppermost leaves are markedly smaller than the others. The adult plant bears 5-7 closely grouped leaves atop an elongate stem internode 0.9–1.01 m. long. The leaves are distichous but appear to be spirally arranged through the twisting of the stem. Rhizome to 1.9 cm. long. *Cataphylls* (1)2–3, thin, pliable, membranaceous above, basal portion fleshier, semicoriaceous, narrowly ovate to linear-ovate, obtuse to acute, mucronulate, dark green or clouded green purple, glabrous, occasionally outer 1 or 2 subglabrous with microscopic hairs; within, shiny green, glabrous, 7–40 cm. long, the uppermost ones 31–40 cm. long, remaining live and green throughout flowering. *Leaf blade* suboblique, narrowly elliptical, apex acuminate, base unequal, acute, 23.2–34 cm. long and 6.4–10.5 cm. wide. Leaf surface above shiny deep grass green, midrib yellow green, glabrous except for distal half of midrib puberulent, lower surface grey green opaque, glabrous, midrib yellow green. *Pulvinus* olive green puberulent on upper side and at the basal junction with the leaf sheath, the rest glabrous, 0.6–1.9 cm. long. *Petiole* absent or extremely rare. *Leaf sheath* stiff, parchment like, not auriculate, green, toward base dark greenish purple, quite dark just above the node, center back for up to 9–10 cm. hispid, margins and rest of the leaf sheath glabrous, 10.7–15 cm. long in subtending leaf, others 14.5–23 cm. long. Nodal area slightly swollen, yellow green. *Stem* green to olive drab, darkest at base, becoming lighter green above, dark green purple just below a node, glabrous, rarely with microscopic hairs on the portion covered by the cataphylls.

Inflorescence terminal, subtended by a foliage leaf, imbricate, cylindric to subovoid, 6.8–8 cm. high and 2.8–4.5 cm. wide. *Peduncle* green, glabrous, 11.3–17 cm. long. *Bracts* 8–11, spiral, imbricate, semicoriaceous, apex spreading, patent to occasionally slightly recurved; depressed ovate to subreniform, apex obtuse with an acumen, shiny grass green, apical portion turning dark purple with age, glabrous; within, pale green, glabrous, 2.5–3 cm. high and 4.3–5.8 cm. wide, each subtending at least 6 flower pairs. *Bicarinate prophyll* membranaceous, rotund, translucent yellow green, with age tinged purple at apex, glabrous, *ca.* 2.4–2.7 cm. long and *ca.* 2.4–2.5 cm. wide. *Secondary bracts* membranaceous, broadly elliptic, apex truncate, translucent yellow green, glabrous, *ca.* 2.35 cm. high and *ca.* 2.2 cm. broad. *Bracteoles* subtending individual flowers, membranaceous, pale yellow green, glabrous *ca.* 2.3-2.4 cm. long and *ca.* 0.3–0.4 cm. wide.

Flowers never opening spontaneously. *Sepals* channeled, margins membranaceous, center and base semispongy, narrowly oblong, obtuse, marginal edges inrolled, appearing acute, basal 1/3 of sepal white, upper 2/3 chartreuse green, glabrous, 3.1–3.25 cm. long and 0.9–0.95 cm. wide. *Corolla tube* white, glabrous, 3.8–4 cm. long; corolla lobes unequal, narrowly oblong-elliptic, obtuse, cream white tinged faint yellow green at apex, 1.5–1.6 cm. long and 0.55–0.83 cm. wide. *Outer staminode* obtriangular, tridentate, white, 1-1.1 cm. long and ca. 0.6 cm. wide. *Callose staminode* rectangular, apex truncate, irregularly bidentate, pale golden, 1-1.1 cm. long. *Cucullate staminode* cream colored at apex, golden to pale orange below, *ca.* 0.9 cm. long and *ca.* 0.5 cm. wide, bearing a fingerlike channeled appendage subterminally. *Filament* cream to yellow orange, bearing a reduced lateral appendage to 1.5 mm. wide; anther 3 mm. long. Style and stigma golden. *Ovary* cream white, glabrous, 3 mm. high. *Capsule* thin, smooth, obovoid, rounded at the apex, yellow green to yellowish with tinge of purple at apex, 1.25-1.5 cm. high and 0.95-1.05 cm. wide, crowned by a purple to purple green expanded, live calyx, sepals oblong, 3.5-3.9 cm. long and 1.2-1.25 cm. wide. *Seeds* usually 3 per capsule, trigonous, slate grey, rugose on outer surface, 0.9-1.05 cm. high and 0.5-0.6 cm. wide, bearing a basal white aril 3-4 mm. high.

Specimens seen: PANAMÁ. Colón: Río Guanche, mature forest about 1.5 mi. upstream from bridge, *ca.* 10 m., flowered in cultivation, 27 Sept. 1971, *H. Kennedy 1158* (K).

COSTA RICA. Limón: La Lola Cacao Institute, near Milla 28, within cacao plantation and adjacent secondary forest, 29 Feb. 1972, *H. Kennedy and R. L. Dressler 1389A* (DUKE); primary forest, Alto Guayacan, S.E. of Siquirres, 700 m., 12 Oct. 1967, *R. W. Lent 1409* (F).

Flowering: *Calathea venusta* flowers during the late rainy season, September through December.

Pollinator: *Calathea venusta* was observed to be pollinated by *Eulaema speciosa* (Mocsary) during November at the type locality.

Distribution and habitat: *Calathea venusta* is known only from the Atlantic lowlands. It is known from a single locality near Portobelo, Panamá, and from two localities in Limón Province, Costa Rica; the northernmost locality is near Siquirres. The vegetation in these areas is tropical wet (evergreen) forest; the Portobelo area is very wet, verging on the tropical rain forest of Beard (1944). Whether this species is restricted to the wet forest habitat is uncertain. Certainly it is not common in the slightly less wet forests near the coast in Bocas del Toro, Panamá, or it would be better represented in collections. *Calathea venusta* is decidedly shade tolerant and occurs within the forest on moderate (apparently well-drained) slopes and often on the upper banks of stream channels.

Variation: No obvious variation in the qualitative characters of *Calathea venusta* was observed in the three collections.

Relationship: *Calathea venusta* is most closely related to *Calathea donnell-smithii.* The bracts of both have a similar texture, shape, color, and color change. Their leaves are strictly cauline and juveniles have relatively numerous leaves. The leaf texture and venation are similar. *Calathea venusta* is somewhat more distantly related to *C. foliosa,* sharing with *C. foliosa* the spiral disposition of leaves.

Notes: The seeds of *Calathea venusta* germinate during the latter part of the rainy season.

12. *Calathea donnell-smithii* K. Schumann

(Plate 17, Figure 16)

Calathea donnell-smithii Schum., Pflanzenreich IV. 48: 75. 1902. Lectotype (here chosen): Haies à Turrialba, 570 m., Prov. Cartago, Costa Rica, Nov. 1893, *A. O. Tonduz 8368* (BR!; isolectotypes CR!, P!).
Calathea picta sensu Woodson *non* Hook. f., Ann. Missouri Bot. Gard. 29: 334. 1942.

Caulescent herb 0.8-1.85 m. high. Leaves up to 9 in juvenile plants, the uppermost leaves smaller than the mid and lower leaves. Leaves 4-6(7) in adults, none basal, distichous, borne above a long stem internode of 35-85.5 cm. in length. Roots with root-tubers, these common in young plants, few to absent in mature plants. *Cataphylls* usually 2-3, narrowly ovate, acute, mucronulate, green, frequently tinged purple at base, glabrous; within, shiny light green, glabrous, innermost 18.9-46 cm. long. *Leaf blade* pliable, papyraceous, ovate, acuminate to attenuate-acuminate, base rounded to subtruncate, shortly and broadly acuminate at junction to pulvinus, 14.7-30 cm. long and 6-12.5 cm. wide. Leaf opaque above, semishiny in younger leaves, deep green, glabrous, except for puberulent midrib; below, pale dull grey green, often tinged with purple, glabrous or with puberulent veins (14X), midrib yellow green, puberulent. *Pulvinus* olive green, glabrous except for row of hairs above, 0.2-1.4 cm. long. *Petiole* absent. *Leaf sheath* auriculate at apex, papyraceous, green, glabrous, darker green at base; subtending leaf 5.8-10.8 cm. long, others 10.5-37.5 cm. long. *Stem* dark green, glabrous.

Inflorescence terminal, ellipsoid to cylindric, 4.4-8.5 cm. high and (1.5)2-4.7 cm. in diameter. *Peduncle* green or green streaked with purple, glabrous, 7.8-17.5(19.5) cm. long. *Bracts* imbricate, 4-11, depressed-ovate, obtuse to rounded with short acumen at center, green when young, with age becoming entirely dark purple, glabrous or minutely pubescent in basal 2-3 mm. at junction with peduncle; within, glabrous, (1.6)1.8-2.5 cm. high and 3-4.1 cm. wide, each subtending 6 or more flower pairs. *Bicarinate prophyll* membranaceous, broadly ovate, obtuse, translucent pale yellow green, glabrous, 1.5-2 cm. high and 1.7-2 cm. wide. *Bracteoles* subtending individual flowers translucent pale yellow green, 1.6-2.2 cm. long.

Flower never opens spontaneously. *Sepals* channeled, semispongy centrally and basally, margins herbaceous, narrowly obovate, acute, basal 1/3 white, upper 2/3 green, glabrous, 2.4-3.6 cm. long and 0.85-1 cm. wide. *Corolla tube* cream, glabrous, 3.7-4.3 cm. long, occasionally with a staminal tube to 3.8 mm.; lobes obovate to elliptic-oblong, obtuse to 90°, margins inrolled appearing acute, cream to faint greenish yellow basally, apex tinged greenish, 1.4-1.8(2) cm. long and 0.7-1 cm. wide. *Outer staminode* obovate, bilobed, rounded or rounded with acumen, pale cream white, 1-1.3 cm. long and 0.5-0.9 cm. wide. *Callose staminode* rectangular, truncate with an acumen centrally, cream yellow, 1-1.15 cm. long. *Cucullate staminode* cream yellow at tip, base golden, 0.9-1 cm. long and 0.45-0.55 cm. in width, bearing a fingerlike channeled appendage *ca.* 1.5 mm. long at the apex. *Filament* bearing narrow lateral petaloid appendage to 1.5 mm. wide; anther free or petaloid appendage joined to basal 1/3, cream, to 3.5 mm. long. Style and stigma golden. *Ovary* cream, glabrous, 2-3 mm. in height. *Capsule* trigonal, obovoid, 1.1-1.3 cm. high and 0.9-1.1 cm. in diameter, crowned by a live, enlarged calyx, sepals to 1.1 cm. wide at base. Sepals with age turning dark purple on capsule. *Seeds* usually 3 per capsule, trigonous, rugose on outer surface, slate grey, 6.2-8.0 mm. long and 4.3-6.0 mm. in diameter; bearing a basal white aril *ca.* 3 mm. high.

Representative specimens seen: PANAMÁ. Panamá: summit of Cerro Campana, 800-1000 m., 1 Sept. 1940, *P. H. Allen 2219* (MO); Cerro Campana, 800 m., 7 Jan. 1947, *P. H. Allen 3964* (MO); along highway near top of Cerro Campana, 10 Sept. 1970, *T. Croat 12081* (MO); region of Cerro Azul, Finca El Indio, *ca.* 750 m., 29 Aug. 1972, *H. Kennedy 1521* (F); Cerro Campana *ca.* 1 mi. past the Florida State field station, *ca.* 800 m., 3 Sept. 1972, *H. Kennedy 1539* (MO); Cerro Campana, cloud forest, 10 Dec. 1967, *W. H. Lewis et al. 3041* (MO, UC); summit of Cerro Campana, rain forest, 31 Mar. 1969, *D. M. Porter et al. 4971* (MO); W. slope of Cerro Campana, 2500 ft., 9 Nov. 1965, *E. L. Tyson et al. 2389* (MO). Coclé: trail to Las Minas N. of El Valle de Antón, 1000 m., 10 May 1941, *P. H. Allen 2459* (US); hills N. of El Valle, vicinity of El Valle at edge of road above valley, 29 Aug. 1960, *J. E. Ebinger 1117* (GH, MO).

COSTA RICA. Puntarenas: Osa Peninsula, flood plain of Río Sandalo, 1-10 m., 24 Aug. 1936, *C. W. Dodge and V. F. Goerger 10145* (CR, F); entre Agua Buena y San Vito de Java, 1200 m., 22 Oct. 1964, *A. Jiménez M. 2477* (CR, F, NY); Osa Peninsula, 2.5 mi. S.W. of Rincón, 8° 42′ N. x 83° 29′ W., forest across from field station, 15-30 m., 10 Sept. 1972, *H. Kennedy 1604* (CR, F, K, MO, U); San Vito de Java, Las Cruces, 1300 m., 14 Sept. 1972, *H. Kennedy 1647* (BM, DAV, DUKE, MO); forest and forest edges on and around Wilson's Finca, 6 km. S. of San Vito de Java, *ca.* 4000 ft., 18 Aug. 1967, *P. Raven 21935* (F); 19 Aug. 1967, *P. Raven 21942* (F). San José: remnant evergreen forest along the river and on very steep slopes above the river Río Chirripo del Pacífico, between Canaán and Chimirol, at 1000 m., (General Valley) 9° 27′ N. x 83° 37′ W., 30 Dec. 1969, *W. C. Burger and R. L. Leisner 7109* (CR, F); vicinity of El General, 850 m., July 1936, *A. F. Skutch 2778* (GH, K, MICH, MO, NY, S, US). Cartago: Hacienda La Gloria, Juan Viña, Dec. 1960, *O. Jiménez s.n.* (CR); Turrialba Agricultural Station, 600-700 m., 25 July 1968, *H. Kennedy 57* (F); Turrialba, forest below Agricultural Institute, 600-700 m., 12 Oct. 1970, *H. Kennedy 640* (DAV, F); sabanillas de Tucurrique, 800 m., 20 May 1941, *J. León 751* (CR, F); vicinity of Pejibaye *ca.* 900 m., 7-8 Feb. 1926, *P. C. Standley and J. Valerio 46765* (US); forêts de Tuis, 650 m., *A. O. Tonduz 11384* (CR, US), cited in the original description. Alajuela: Colinas de San Pedro de San Ramón, 1050-1075 m., 18 Sept. 1925, *A. M. Brenes 223, Brenes Herb. 4438* (CR, F); colinas de San Pedro de San Ramón, bois, 1075 m., 17 Oct. 1925, *A. M. Brenes 326* (CR); vicinity of Capulín, on Río Grande de Tárcoles, *ca.* 80 m., 2 Apr. 1924, *P. C. Standley 40179* (US). Province unknown: bei Aguacate, Aug. 1857, *C. Hoffmann 762* (destroyed in B, photo F), cited in original description, date and locality *fide* Schumann; Ad Faris, Nov. 1847, *Ørsted 1* (C).

NICARAGUA. Dept. unknown: near Tortuga, May 1847, *A. S. Ørsted s.n.* (C) mixed collection of *C. donnell-smithii* and *C. marantifolia.*

Flowering: *Calathea donnell-smithii* flowers during the rainy season, from July through December, peaking in September and November. Collections of flowering individuals have been made as late as

February and March. Flowering may continue early into the dry season in areas where the soil moisture remains sufficiently high. A single individual was found in flower in April near San Vito de Java, Costa Rica, in a watered garden.

Pollinator: At Cerro Jefe, Panamá Province, Panamá, *Calathea donnell-smithii* was observed to be pollinated by *Eulaema speciosa* (Mocsary) and less commonly by *Eulaema meriana* (Olivier). *Euglossa asarophora* Moure was observed to rob the flower at the same locality.

Distribution and habitat: The southern- and easternmost locality for *Calathea donnell-smithii* is the region of Cerro Jefe, Panamá Province. In Panamá it occurs between 700 and 1000 m. in premontane rain forest and is prevalent on Cerro Campana and again on the slopes above El Valle. In southern Costa Rica it occurs on the Pacific slope in premontane wet forest above 800 m. near San Vito de Java and El General, but also occurs near sea level on the Osa Peninsula. Farther north, it occurs on the Atlantic slope in premontane wet forest near Turrialba and Juan Viñas. The northernmost known locality is near San Ramón, Costa Rica, at 1000 m., a wetter climate. Since so few collections have been made on the Atlantic slope in Panamá, the consistent appearance of this species on the Pacific side may be an artifact of collection. *Calathea donnell-smithii* occurs scattered within the forest in moderate shade; however, it appears to do well or better near forest edges such as along road cuts or around tree fall areas. It is not found in open, fully exposed areas.

Variation: There is no very striking geographical variation in *Calathea donnell-smithii* throughout its range. The Osa Peninsula, Costa Rica, population is the most noticeable variant. The leaf in this population is elliptic with a rounded apex bearing a short acumen. The ratio of width to length of this population is about 1:2; in other populations it is about 2:5. The leaves of individuals from San Vito, Costa Rica, have an acuminate apex, others grade from acuminate to obtuse with an acuminate point. In the majority of populations, the leaves are glabrous underneath with minutely puberulent veins. In the Cerro Campana population and in most individuals from Turrialba, the veins also are glabrous. There is a tendency for the Panamanian populations to have minute puberulence at the junction of the bract with peduncle, whereas in most Costa Rican plants (except Osa) the bracts are entirely glabrous. One individual (*Kennedy 640A*, F) from Turrialba has markedly larger leaves that are densely puberulent below. The leaf apex is obtuse with a short acumen and the base is subtruncate. Other plants in the same area (*Kennedy 640*, F, DAV) have a glabrous leaf and obtuse base. These were growing sympatrically with *Calathea marantifolia* Standley whose leaf shape and pubescence are more similar to that of the aberrant individual, suggesting a possible hybrid origin of the variant plants.

Relationship: *Calathea donnell-smithii* is most closely related to *C. venusta* and only slightly less so to *C. longiflora*. All three share a similar inflorescence morphology. The shape, color, and texture of the bracts are quite similar. However, the bracts are slightly recurved in *C. longiflora* and *C. venusta*, whereas they are not so in *C. donnell-smithii*. Both *C. venusta* and *C. donnell-smithii* lack basal leaves, but *C. longiflora* often has a basal leaf. More distantly related are *C. foliosa*, with similarities of habit and leaf structure, and *C. marantifolia*, with a similar bract texture and shape.

Notes: The bracts of *Calathea donnell-smithii* are frequently attacked by the larvae of leaf-mining beetles, hispids (Hispinae, Chrysomelidae), which leave vertical marks on the bracts. This type of damage is almost diagnostic for this species because of its prevalence. This damage is evident in the photograph of *Hoffmann 762* (F).

In the *Flora of Panama* (Woodson and Schery, 1945), *Calathea donnell-smithii* is listed as *C. picta* Hook. f. *Calathea picta*, however, is a synonym of *C. warscewiczii* and is quite different (Kennedy, 1973). The name *Calathea violacea* listed in the *Flora of Costa Rica* (Standley, 1937) represents a composite misidentification. Standley identified *C. donnell-smithii* (*Standley and Valerio 46765*, US) as *C. violacea*, as well as a specimen of *C. warscewiczii*. This misnomer was probably the result of the presence of the purple bracts which characterize old inflorescences of *C. warscewiczii* and *C. donnell-smithii*. *Calathea violacea* has green bracts and open, purple flowers.

13. **Calathea longiflora** H. Kennedy, sp. nov.

(Plate 18, Figure 12)

Planta humilis vix metralis. Folia breviter petiolata, petiolus tota longitudine callosus vel petioli pars superior ad 3 cm. longa glabra supra puberula, ovata vel anguste ovata acuminata vel caudata basi rotundata et breviter acuminata mediano et venis minutissime puberulentis ceterum utrinque glabra, chartacea sulcata smaragdina; vagina supra tomentosa subtus subglabra. Spica imbricata ovata vel subcylindrica complanata folia comitata pedunculo gracili minutissime puberulento sustenta; bracteae paucae spiraliter depresse ovatae acuminatae apice patenti semicoriaceae supra glabrae subtus puberulentae smaragdinae postea purpureae; paria florum ultra 5 mesophyllis et bracteolis membranaceis comitata; ovarium glabrum; sepala anguste obovato-elliptica viridia; corollae tubus albidus lobi anguste obovato-elliptici cremei; staminodium exterius obovatum unguiculatum irregulare; callosum rectangulare; cucullatum cremeum subtus aureum. Capsula calyce demum coronata; semina rugosa trigona dorso rotundata schistacea, arillo albo munita.

Caulescent herb 0.45-1.1 m. high. Each shoot bearing 2-4 leaves, 0-1 basal and 2-3 cauline. Roots bearing root-tubers. *Cataphylls* 2-3, narrowly ovate, apex acute, mucronulate, green, puberulent, innermost 25-38 cm. long, next 8-15 cm. long, the lowest ones frequently rotting away. *Leaf blade* chartaceous, major veins—frequently every other vein—raised, semishiny, thicker than the surrounding mesophyll, the minor veins lower, dull, not appreciably thicker than surrounding mesophyll, giving the leaf a rippled or finely pleated appearance; ovate to narrowly ovate, apex attenuate-acuminate to caudate, base rounded to truncate shortly acuminate at the junction with the pulvinus, 13-26.2 cm. long and 6-11.3 cm. wide in subtending leaf; others 21.3-39.6 cm. long and 7.9-14.1 cm. wide. Upper surface grass green with slight sheen in younger leaves, with age, dulling, opaque, glabrous with puberulent midrib, more pronounced in distal half. The lower surface light grey green, midrib and veins minutely puberulent. *Pulvinus* glabrous except for a row of hairs along the upper surface, olive green, lighter than petiole, base extended along back and sides of petiole, slightly wider than petiole, junction abrupt, distinct, 0.6-3 cm. long. *Petiole* absent on subtending leaf and generally absent on the next lower leaf also if 3 or more leaves are present. Petiole green, subglabrous to tomentose increasingly so basally, bearing a narrow groove *ca.* 2 mm. wide along the upper surface lined with fine hairs, 0-17.5 cm. long. *Leaf sheath* densely ciliate along margins especially apical portion, the back and sides glabrous becoming tomentose a few centimeters above the node in the upper leaf or leaves, 3rd or 4th leaf sheath tomentose above gradually becoming subglabrate toward base, the base at the junction with stem pubescent; 4-7.1 cm. long in subtending leaf, 11-56 cm. long in lower ones. Nodal area yellow green pubescent. *Stem* internode tomentose below node becoming subglabrous lower down. First internode below subtending leaf 7.5-29 cm.; second internode 27.5-62 cm.

Inflorescence terminal, subtended by a leaf. *Peduncle* green turning purple with age, frequently tinged purple just below the inflorescence, minutely scattered puberulent, not visible to the naked eye, 5.5-7.5 cm. long. Inflorescence sublax, imbricate, ovoid, 4-5 cm. high and 2.5-3.5 cm. wide. *Bracts* spiral 5-9, depressed ovate to reniform, apex rounded abruptly acuminate at center, acumen *ca.* 4 mm., semicoriaceous, basal portion imbricate, apex recurved to patent, pale green minutely tomentose at base, upper portion glabrous smooth shiny green, recurved apex turning purple with age; inner surface glabrous, 2.3-2.8 cm. high and 3.1-4.5 cm. wide; each subtending 5 or more flower pairs. *Bicarinate prophyll* membranaceous broadly elliptic to subrotund, obtuse to rounded, translucent faint green at base, semiopaque grass green above, glabrous, 1.7-2 cm. high, largest in lower bracts decreasing slightly in upper bracts, and (1.3)1.8-2 cm. wide. *Secondary bracts* membranaceous, broadly elliptic to broadly obovate, rounded to subtruncate, translucent faint green at base, grass green above, glabrous, 1.6-1.9 cm. high and 1.6-1.7 cm. wide. *Bracteoles* subtending individual flowers usually 2 per flower pair, translucent light green, glabrous, 1.8-1.9 cm. high.

Flower never opens spontaneously. *Sepals* narrowly obovate to narrowly obovate-elliptic, acute, basal 1/2 white, distal 1/2 light green, herbaceous to semispongy at base, glabrous 2.4-2.8 cm. high and 0.4-0.8 cm. wide. *Corolla tube* white, glabrous, 3.6-4.1 cm. long; corolla lobes unequal, narrowly obovate to obovate-elliptic, obtuse, cream white at base, pale yellowish above, apex tinged yellow green, (1.1)1.3-1.5 cm. long and 0.55-0.9 cm. wide. *Outer staminode* obovate, apex rounded, irregular, forming a shallow cup, pale cream, 0.9-1.15 cm. long and 0.4-0.7 cm. wide. *Callose staminode* rectangular, truncate with small acumen centrally, cream colored 0.9-1.1 cm. high. *Cucullate staminode* cream colored at apex, centrally and basally golden, 0.8-0.95 cm. long and 0.4-0.45 cm. wide, provided with a small curved appendage at the tip, *ca.* 1.5 mm. high. *Filament* with a narrow lateral appendage to 1.5 mm. wide; anther 2.5-3 mm. long. Style and stigma golden. *Ovary* white, glabrous 2-2.5 mm. high. *Capsule* pale green tinged purple at apex, 1.1-1.3 cm. high and 0.8-0.95 cm. wide, crowned by a persistent, live calyx, sepals becoming purple with age. *Seeds* usually 3 per capsule, trigonous, rugose on the outer surface, slate grey, 5.5-6.5 mm. high and 3.5-5 mm. wide, provided with a basal white aril *ca.* 3 mm. long.

Type: COSTA RICA. Prov. Puntarenas: Osa Peninsula, 2.5 mi. S.W. of Rincón, Quebrada Agua Buena, 15-30 m., 8° 42′ N. x 83° 29′ W., 10 Sept. 1972, *H. Kennedy 1601* (holotype F 1751757!; isotypes CR!, DAV!, F!, GH!, K!, MO!, U!, US!).

Specimens seen: COSTA RICA. Puntarenas: Osa Peninsula, *ca.* 5 km. W. of Rincón, moist forest in valley bottoms, forest with open understory on steep slopes and ridges and areas of secondary vegetation, N. and W. of the airfield, 50-200 m., 8° 42′ N. x 83° 31′ W., 9-12 Jan. 1970, *W. C. Burger and R. L. Liesner 7192* (F); *W. C. Burger and R. L. Liesner 7226* (F); 2.5 mi. S.W. of Rincón, 15-30 m., 8° 42′ N. x 83° 29′ W., 11 Aug. 1967, *H. Kennedy 41* (F); Quebrada Agua Buena near T. S. C. field station, 30-50 m., 8° 42′ N. x 83° 29′ W., 19 Oct. 1960, *H. Kennedy 647* (BM, DAV, MO, US); 30-50 m., 4 Sept. 1971, *H. Kennedy 1142A* (F); across the Río Agua Buena from T. S. C. field station, 15-30 m., 8° 42′ N. x 83° 29′ W., 11 Nov. 1972, *H. Kennedy 1936* (MO); 16 Nov. 1972, *H. Kennedy 1998* (MO); 4 mi. W. of Rincón de Osa, deep forest near the airport area, 100 ft., 11 Aug. 1967, *P. H. Raven 21720* (F).

Flowering: *Calathea longiflora* flowers during the rainy season, July through December, with peak flowering primarily in the latter part, September through November. It has been collected in flower as late as 9 January.

Pollinator: At the type locality, *Calathea longiflora* was observed to be pollinated by *Euglossa asarophora* Moure during November.

Distribution and habitat: At present *Calathea longiflora* is known only from the type locality and the area some 2-3 km. downstream along the Río Agua Buena, Osa Peninsula, Costa Rica. The vegetation is wet (evergreen) forest. This species occurs in moderate shade within the forest, especially near stream banks, washes, or runoff channels.

Variation: The geographic variation of *Calathea longiflora* is unknown; the Osa Peninsula, Costa Rica, population is quite uniform in qualitative characters. The variation in habit between individuals is of interest as it effectively bridges the gap between the species with only cauline leaves and those with the common arrangement of basal plus cauline leaves. Frequently all the shoots of an individual will show the same habit but younger (smaller) shoots more commonly have a basal leaf.

Relationship: *Calathea longiflora* most closely resembles *Calathea donnell-smithii,* in habit, form of inflorescence, and the characteristic color change with age of the bracts and sepals from green to dark purple. *Calathea longiflora* differs from *C. donnell-smithii* in its patent bract margins, smaller stature, occasional presence of a basal leaf, and distinctive rippled leaf surface. The rippled leaf character effectively distinguishes *C. longiflora* from all other Costa Rican and Panamanian species in the series *Scapifoliae.* The one exception is an undescribed species from near San Ramón, Costa Rica, which has a more deeply grooved leaf and differs from *C. longiflora* in habit and leaf shape.

Notes: Six artificial self-pollinations of *Calathea longiflora* from the Osa Peninsula, Costa Rica, under cultivation in the Canal Zone failed to produce seed. In some of the pollinations the capsule began to develop but aborted before maturity.

14. *Calathea leucostachys* Hooker f.

(Plate 19, Figure 17)

Calathea leucostachys Hook. f., Bot. Mag. 101: *t. 6205.* 1875. Type: Hort. Veitch, 24 Oct. 1874, *Enders s.n.* (K!). The specimen is of a plant sent by Mr. Enders from Costa Rica which "flowered in Mr. Veitch's establishment in Chelsea in Oct. 1874" (Hooker, 1875).

Calathea valeriana Standl. Field Mus. Publ. Bot. 18: 194. 1937. Type: Forests between La Balsa and Cataratas de San Ramón, 850 m., Alajuela Prov., Costa Rica, Oct. 1925, *Brenes 316A* = *Brenes Herbarium 4531A* (F 852624!; isotype CR 18401!).

Herb 0.3-0.7 m. high. Shoots bear 5-9 (or more) leaves. Internodes reduced, the leaves distributed in a more or less rosette fashion, the internode between the uppermost leaf subtending the inflorescence and the second leaf elongate (8.5-20 cm.) in relation to the other internodes (about 1 cm. long). The lower leaves gradually rot off, leaving a stem portion which then roots, occasionally rooting through the bases of the lower extant leaves. This above-ground stem portion may be as much as 24 cm. in length. The stem is frequently held above ground level by the stiltlike roots. Additional branch shoots generally arise in the axils of the extant leaves rather than from the rhizome. *Cataphylls* narrowly ovate, semifleshy toward the base, acute, apiculate, outer surface green, villous; inner surface glabrous, short-lived, absent in

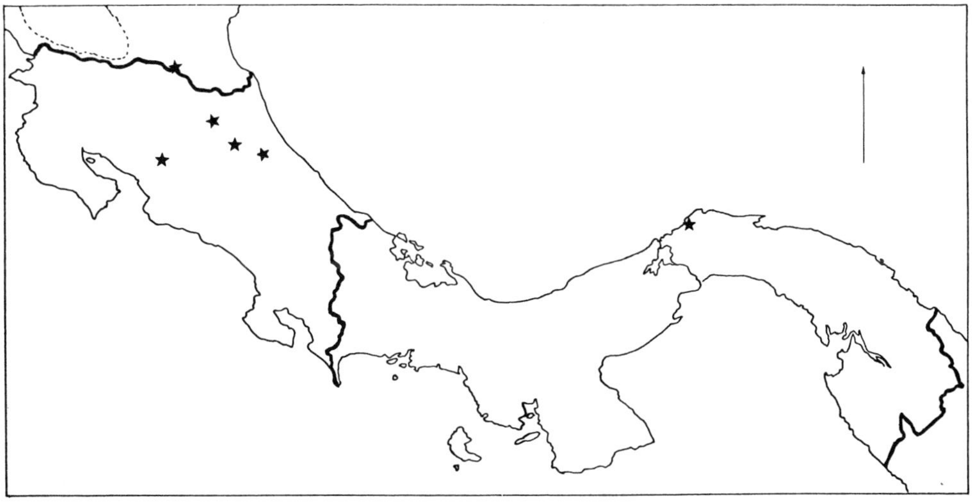

Fig. 17. Distribution of *Calathea leucostachys* in Costa Rica and Panamá.

flowering individuals, to 17.5 cm. high. *Leaf blade* papyraceous, pliable, elliptic to oblong, occasionally subobovate, obtuse to rounded with a point, base subequal, rounded, 23-41 cm. long and 11-21 cm. wide. Leaf surface above opaque grass green to dark green, villous, midrib yellow green, light green fishtail pattern to 2.5 cm. wide along the midrib may or may not be present (this pattern is more prevalent in younger leaves); leaf under surface deep red purple or dull grey green with yellow green midrib, glabrous except puberulent midrib to villous throughout. *Pulvinus* olive green or reddish purple, villous, 1.2-3 cm. long. *Petiole* green, villous, 0-0.7(1.5) cm. long, almost always absent. *Leaf sheath* semichartaceous, green, villous above especially along the back, margins and lower 1/3 subvillous to puberulent; shiny, green, glabrous within, 14-26 cm. long in subtending leaf, others 24-41 cm. long. *Stem* green, subvillous.

Inflorescence terminal, usually subtended by a foliage leaf. *Peduncle* light green, villous, 16-26 cm. long. Inflorescence imbricate semilax, ovoid to cylindric, 11-12.2 cm. high and 7-11 cm. wide, bract tip to bract tip, live measurement. *Bracts* (7)10-22, lower bracts transverse broadly ovate to ovate in upper bracts, apex of lower bracts often retuse with acute tip in the sinus changing in upper bracts to rounded or rounded with an acumen centrally, soft pliable, leathery, (3.4)4-7.3 cm. high and 4.1-6 cm. wide, marginal portion hemispherical to oblong, patent to reflexed. Basal imbricate portion abaxially light tannish, villous; adaxially cream, shiny, glabrous. Abaxially, marginal portion white, villous; with age this becomes tinged pink purple and finally deep red purple. Rachis of inflorescence pale tan, densely villous. Each bract subtends up to 5(6) flower pairs. *Bicarinate prophyll* membranaceous, obovate to broadly obovate, obtuse, translucent white to pale tan, puberulent above, subglabrous below, hairs golden tan, 2.15-3.25 cm. high and 1.15-1.7 cm. wide carina to carina, total width 2.2-2.4 cm. *Secondary bracts* membranaceous, obovate-elliptic, apex obtuse to subrounded, translucent white to pale tan, upper 2/3 puberulent, 2.2-3.2 cm. high and 2-2.5 cm. wide. *Bracteoles* subtending individual flowers membranaceous, glabrous, transparent, (1.9)2.2-2.9 cm. high.

Flower white, never opening spontaneously. *Sepals* channeled, thicker semispongy at base, narrowly obovate, obtuse, white, semishiny, pilose above, subpilose below, 2.5-3.6 cm. long and 0.55-0.9 cm. wide. *Corolla tube* white, glabrous, 2.5-3.6 cm. long; lobes unequal, obovate to subelliptic, apex obtuse to 90°, white, glabrous, 1.3-1.55 cm. long and 0.55-0.9 cm. wide. *Outer staminode* highly reduced, often absent, narrowly obovate to narrowly obtrullate, apex dentate, white, up to 0.3 cm. long and 0.2 cm. wide. *Callose staminode* rectangular, truncate with an acumen centrally, white at tip, faintly tinged yellow orange in throat, 0.9-1.1 cm. long. *Cucullate staminode* white, toward base cream, *ca.* 0.95 cm. long and 0.45-0.5 cm. wide, bearing a short channeled appendage at tip to 1.5 mm. long. *Filament* with narrow lateral appendage, white; anther cream yellow to 2.5 mm. in length. Style and stigma golden orange.

Ovary white, pubescent or glabrous below with a ring of hairs at apex, *ca.* 2.5 mm. high. *Capsule* trigonous, obovoid, cream colored, 1.3–1.5 cm. high and 0.85–1 cm. wide, crowned by a persistent live calyx, sepals oblong enlarged to 1.4 cm. wide, the apices of the sepals spreading to subpatent. *Seeds* usually 3 per capsule, trigonous, rugose on outer surface, slate grey 8.5–9.5 mm. high and 4.5–5.5 mm. in diameter; bearing a basal white aril, 3.5–4 mm. high.

Specimens seen: PANAMÁ. Colón: Río Guanche, *ca.* 1.5 mi. upstream from the bridge, 17 July 1971, *H. Kennedy, R. Foster, R. L. Dressler 1157* (DAV); Río Guanche, *ca.* 1.5 mi. upstream from the bridge, 10 m., 23 Aug. 1972, *H. Kennedy and R. L. Dressler 1510* (F).

COSTA RICA. Limón: en bosque vecino al puente del Río Toro Amarillo, 5.5 km. al S.O. de Guapiles, 300 m., 9 May 1963, *A. Jiménez M. 1025* (F); en bosque cercano al Río Toro Amarillo, 6 km. al Sur Oeste de Guapiles, 400 m., 18 July 1964, *A. Jiménez M. 2112* (BM, F, GH); Siquirres, cultivated at Wilson's Finca, Las Cruces, near San Vito, flowered in cultivation 23 Sept. 1969, *H. Kennedy 385* (F); Siquirres, flowered in cultivation, 14 Sept. 1972, *H. Kennedy 1648* (CR, DUKE, F, K, MO, U, US); Toro Amarillo, 200 m., 10° 23′ N. x 84° 15′ W., 18 July 1964, *M. F. Tessene 1409* (F). Alajuela: bois, alto de Cataratas de San Ramón, 750–800 m., 5 Sept. 1925, *A. M. Brenes 211, Brenes Herb. 4426* (CR); San Ramón, premontane forest, 15–20 km. N.N.W. of San Ramón along Calle Los Angeles, beyond Balsa, *ca.* 1000 m., 19 Sept. 1972, *H. Kennedy 1661* (F); near shore of Laguna Hule, 750 m., 6 Aug. 1967, *R. W. Lent 1194* (F); forested slopes and roadside about 19 km. N. of San Ramón, *ca.* 3400 ft., 16 Jan. 1968, *R. L. Wilbur and D. E. Stone 9700* (DUKE).

NICARAGUA. Río San Juan: in tall forest near Río San Juan at "El Relos" *ca.* midpoint between El Castillo and Delta de San Juan, 0–50 m., 23 Mar. 1961, *G. S. Bunting and L. Licht 757* (F).

Flowering: *Calathea leucostachys* flowers primarily in the rainy season, July to October. This species may have two flowering periods or in wetter areas the flowering may extend into the dry season.

Distribution and habitat: *Calathea leucostachys* is known mainly from the Atlantic lowlands, but also occurs at 1000 m. near San Ramón, Costa Rica. The easternmost and southernmost locality is the wet (evergreen) tropical forest near Portobelo, Panamá. Siquirres and Toro Amarillo, Costa Rica, and the region around the Río San Juan, Nicaragua, the northernmost locality of *C. leucostachys*, likewise support wet (evergreen) forests. The vegetation around San Ramón is premontane wet or cloud forest. *Calathea leucostachys* usually occurs within the forest, growing in moderate to deep shade. The geographic distribution of this species is somewhat similar to that of *Calathea venusta*.

Variation: Individuals of *Calathea leucostachys* having characters that are most consistent with the type description and plate (Bot. Mag. *t. 6205*) are found in the area around La Balsa, near San Ramón, Costa Rica (*Kennedy 1661* and *Brenes 316A* = *Brenes Herb. 4531A*). This is also the type locality of Standley's *Calathea valeriana*. In the San Ramón population, the leaf blade is dark red purple, glabrous below and is dark green, villous above; the ovary is pubescent and the internode between the subtending and the penultimate leaf is short. The lowland populations from Puerto Viejo and Siquirres, Costa Rica, and the Río Guanche, Panamá, are larger and more robust in habit; have wider leaves, the ratio of length to width *ca.* 1.9–2, whereas that of the individuals from San Ramón is *ca.* 2.4; have a longer internode below the subtending leaf and a glabrous ovary. Plants from the Puerto Viejo and the Siquirres region occasionally have a narrow, fishtail, pale green pattern along the midrib of the leaf (more pronounced in younger leaves). The lower leaf surface in the Puerto Viejo population is pale grey green and villous, whereas it is subglabrous in the Siquirres population. The Panamanian individuals have a distinct pattern on the leaf above, the lower surface tinged light to pale red purple, villous. Within a population the shape of the patent to recurved bract "margin" or apical portion may vary considerably. In the original description Hooker described this portion as "much broader than long" (Hooker, 1875). Among the plants cultivated at Las Cruces Botanical Garden, Costa Rica, from Siquirres there were many in which this apical portion was at least 1.5 times as long as broad in addition to those in which it was indeed broader than long. In no other species of the *Scapifoliae* with closed flowers are the bracts so well developed.

Relationship: *Calathea leucostachys* is most closely related to *Calathea warscewiczii*, sharing a similar inflorescence morphology and texture, and the color of the bracts and flowers. In both species the bract margins are recurved and the bracts undergo a color change with age to a reddish purple. Both species appear to be rather distantly related to any other species in the group, but they appear to be more closely related to the *C. vinosa*–*C. indecora* alliance than to *C. marantifolia* or *C. donnell-smithii* and its allied species.

15. *Calathea warscewiczii* (Mathieu) Koernicke

(Plate 20, Figure 9)

Maranta warscewiczii Mathieu, Mathieu Prospectus, May 1853. Type: the plate in Mathieu Prospectus of a plant grown in cultivation in Berlin introduced by J. von Warscewicz from Central America; material cultivated in Berlin, in the Kew Herbarium, is probably from the type plant. *Phrynium warscewiczii* (Math.) Klotzsch in Allgem. Gartenzeitg. 23: 18. 1855. *Calathea warscewiczii* (Math.) Koernicke, in Gartenfl. 7: 87. 1858. *Phyllodes warscewiczii* (Math.) O. Kuntze, Rev. Gen. Pl. 2: 697. 1891.

Maranta picta Bull, Bull. Cat. New Pl. No. 324, p. 6, pl. p. 4. 1898. Type: Hort. Kew, Dec. 28, 1898, *Hooker s.n.* (K!), a plant which Hooker obtained from Bull (Holttum, pers. comm.). *Calathea picta* (Bull) Hooker f., Bot. Mag. 125: *t. 7674.* 1899. *Calathea bullii* K. Schumann, Pflanzenreich IV. 48: 114. 1902. Based on *Maranta picta* Bull.

Caulescent herb (48)57-120(152) cm. high; bearing (0)3-7 leaves in a basal rosette, 2 (rarely 1 or 3) cauline, subtending the inflorescence. These latter are borne upon an elongated stem internode (38)42.5-103 cm. in height. Roots bearing root-tubers, these more evident in juvenile plants. *Cataphylls* usually 2-3, narrowly ovate, acute to *ca.* 90°, stiffish, fleshy toward base, rapidly decaying and very rarely present in flowering individuals, dark red purple, densely velvety tomentose on outer surface, inner surface pale greenish, glabrous; innermost ones 20-61 cm. long. *Leaf blade* ovate, acuminate, base obtuse to rounded, subtending leaves 8-35 cm. long and 4.7-14.6 cm. wide; basal leaves 18-47 cm. long and 11.4-21 cm. wide. Leaf surface above with velvety sheen, dark green with yellow green fishtail pattern along the center up to 5 cm. from the midrib on each side; below, dark red purple, dense velvety tomentose. *Pulvinus* green to darkly tinged red purple, short velvety tomentose, felt rather than seen, 0.9-6.3 cm. long. *Petiole* dark red purple, dense velvety tomentose, 0.6-34 cm. long in basal leaves, usually absent in subtending leaves, but occasionally up to 6.5 cm. long. *Leaf sheath* chartaceous, often auriculate in subtending leaves, dark red purple, velvety tomentose, 4.8-18(24) cm. long in subtending leaves; basal leaves 24-59.5 cm. long.

Inflorescence terminal, imbricate, cylindrical, 3.5-7.1 cm. long and 3-5.2 cm. wide, bract tip to bract tip. *Peduncle* green, velvety puberulent, 5.5-16 cm. in length. Inflorescence bracts (2)4-16, clasping at base, margin patent to recurved, very broadly to depressed-ovate, basal 1 or 2 subreniform, apex obtuse to rounded, lowermost one usually with a short acumen centrally, upper bracts narrower and longer than lower ones, coriaceous, 2.4-3.4 cm. high and 2.9-5.25 cm. wide, each subtending up to 8 flower pairs. *Bracts* velvety tomentose throughout except adaxial basal imbricate portion; patent to recurved marginal portion white, with age tinged red purple becoming dark red purple in very old inflorescences, basal abaxial portion pale straw yellow. *Bicarinate prophyll* membranaceous, broadly obovate-elliptic, obtuse, translucent pale tan, apical margins puberulent, portion between the carina puberulent to glabrous, 2.1-2.3 cm. high and 1.4-1.9 cm. wide. *Secondary bracts* membranaceous, translucent pale tan, inner ones almost transparent, puberulent at apex to subglabrous, outermost 2.1-2.25 cm. high and 1.8-1.95 cm. wide. *Bracteoles* subtending individual flowers 1 or 2 per flower pair, 1.6-1.8 cm. high and *ca.* 2.5 mm. wide, occasionally highly reduced, glabrous, pale tan.

Flowers never opening spontaneously. *Sepals* narrowly obovate-elliptic, obtuse, channeled, subcoriaceous, base and mid portion slightly spongy, shiny white, glabrous, 2.6-3.3 cm. high and 0.5-0.9 cm. wide. *Corolla tube* white to faint cream, glabrous 2.6-3.5 cm. long; lobes obovate-elliptic, obtuse, marginal edges folded in appearing acute, white to faint cream, 1.2-1.9 cm. long and 0.55-0.85 cm. wide. *Outer staminode* quite variable in shape, narrowly obovate to obtriangular, apex variable, tridentate to rounded or with an acumen, white, 0.9-1.0 cm. long and 0.35-0.5 cm. wide. *Callose staminode* rectangular, fleshy, irregular to subtruncate with acumen centrally, cream colored, 1-1.1 cm. high. *Cucullate staminode* cream white, toward base golden, 0.85-0.9 cm. long and *ca.* 0.4 cm. wide bearing a short fingerlike channeled appendage up to 1.5 mm. long at the tip. *Filament* white, bearing a lateral appendage to 1.5 mm. wide; anther 3 mm. long. Style and stigma golden. *Ovary* white, glabrous, 2-2.5 mm. long. *Capsule* trigonous, ovoid, cream white, 1.2-1.3 cm. high and 0.6-0.9 cm. wide, crowned by a persistent live calyx, sepals enlarged. *Seeds* usually 3 per capsule, trigonous, rugose on outer surface, slate grey, 8-8.5 mm. high and 4-4.5 mm. wide; bearing a basal white aril 2.5-4 mm. high.

Representative specimens seen: PANAMÁ. Bocas del Toro: milla 7.5 on the railroad from Almirante, on the ridge S.W. of the railroad, 17 Nov. 1971, *H. Kennedy and R. L. Dressler 1280* (MO); Almirante, along a stream S.W. of town, 18 Nov. 1971, *H. Kennedy and R. L. Dressler 1284* (BM, BT, DAV, K, MO, NY, P, PMA, S, UC, US); inland (south) from mouth of Changuinola River, banks of river and adjacent rain forest, 10–15 m., 18 Dec. 1966, *W. H. Lewis, J. D. Dwyer, T. S. Elias, K. R. Robertson 982* (MO); Chiriquicito to 5 mi. S. along Río Guarumo, rain forest by river, 5–7 June 1967, *W. H. Lewis et al. 2098* (MO); Rubber Tree Station, 12 Aug. 1923, *H. E. Stork C'119'* (UC, US); Water Valley, 10 Sept. 1940, *H. von Wedel 680* (BM, GH); 23 Sept. 1940, *H. von Wedel 903* (GH, MO, US); vicinity of Chiriquí Lagoon, 8 Oct. 1940, *H. von Wedel 1077* (BM, GH, US); Isla Colón, 21 Oct. 1940, *H. von Wedel 1270* (BM, GH, US); Water Valley, vicinity of Chiriquí Lagoon, 9 Nov. 1940, *H. von Wedel 1586* (MO); Water Valley, 11 Sept. 1941, *H. von Wedel 2655* (GH, MO, US).

COSTA RICA. Limón: Caribbean coastal plain with Cacao plantations, remnants of native wet forest vegetation and littoral vegetation between sea level and 100 m. in the general vicinity of Limón, 3–7 Sept. 1971, *W. C. Burger and M. Burger 8440* (F); vicinity of the U.S. Dept. of Agriculture Rubber Experiment station, Los Diamantes, on Río Santa Clara (1.6 km. E. of Guapiles), 9 July 1949, *ca.* 200 m., *R. W. Holm and H. H. Iltis 345* (A, P); Portete, 12 Feb. 1965, *A. Jiménez M. 2870* (CR, F, NY); La Lola, near mile 28, cacao plantation behind La Lola, trail to Río Madre de Dios, 29 Feb. 1972, *H. Kennedy, R. Andrews and R. L. Dressler 1386* (MO); La Lola, near mile 28, La Lola Cacao Institute, in cacao plantation and adjacent secondary forest, 1 Mar. 1972, *H. Kennedy and R. Andrews 1392* (CR, NY); Carillo, 300 m., 8–9 July 1889, *H. Pittier 1192* (BR, CR, P, US); plantations de bananes Waldeck, prés Madre de Dios, Atlantique, 50 m., Oct. 1896, *H. Pittier 10363* (BR, CR); endivits frais à La Colombiana plaines Atlantiques, llanuras de Sta. Clara, 200 m., June 1899, *H. Pittier 13442* (K, US), one sheet lists *Pittier 7609*; forêts du Río Hondo, 200 m., Aug. 1901, *H. Pittier 16157* (CR); collines de Zent, 31 m., Sept. 1901, *A. Tonduz 14514* (CR); Hacienda de Zent, *United Fruit Co. 246* (US). Cartago: above Río Humo at Pejibaye Farm, 650–800 m., 25 Sept. 1929, *C. W. Dodge and W. S. Thomas 4372* (GH, MO); bois de la valle du Rio Tuís, 600 m., Sept. 1893, *A. Tonduz 8102* (BR); forêts de Tuis, 650 m., Nov. 1897, *A. Tonduz 11382* (CR); a Las Vueltas, Tucurrique, 635 m., Mar. 1899, *A. Tonduz 13299* (CR, US). Alajuela: low tropical rain forest with palms prominent in understory, vicinity of Guatuso de San Raphael (Río Frio) (10° 43′ N. x 84° 48′ W.) 80–100 m., 4 Aug. 1949, *R. H. Holm and H. H. Iltis 892* (A). Heredia: La Selva O.T.S. field station on Río Puerto Viejo, near Pto. Viejo, 6 Aug. 1968, *H. Kennedy 58* (F, US); 15 July 1970, *H. Kennedy 406* (CR, GH, U); Finca La Selva, near Puerto Viejo, 90 m., secondary cacao, 25 Oct. 1971, *P. A. Opler 420* (F). Province unknown: 13 Oct. 1936, *E. Umaña 44* (CR). Limón or Cartago: Pacuare-Brücke, 24 Jan. 1932, *W. Kupper 451* (M).

CULTIVATED SPECIMENS. Kew Gardens, 30 Jan. 1902 (K); Hort. Kew, March 1856, Herb. Hookerianum (K); Mus. Bot. Berol., cult., *K. Koch* (K); Ex horto Bot. Petropolitano 66.5 (LE); Hort. Bot. Vindob., 31 July 1865 (NY); Hort. Bot. Parisiensis, 7 March 1860 (P); April 1856, Hort. Bot. Praga (PR).

Flowering: *Calathea warscewiczii* flowers mainly in the rainy season, though in the Atlantic lowlands the effect of the dry season is not so pronounced and this species often continues flowering into early dry season. Collections of flowering specimens from Bocas del Toro Province, Panamá, have been made from June through December. The flowering time definitely is influenced by the available moisture; in Bocas del Toro Province, *C. warscewiczii* was flowering in early June of 1967, whereas in 1973, when there was a longer than normal dry season, the plants did not even show signs of inflorescence initiation by 18 June. In Limón Province, Costa Rica, collections of flowering plants were made from June through March.

Distribution and habitat: The known distribution of *Calathea warscewiczii* is along the Atlantic slope and lowlands from northwestern Panamá to northern Costa Rica, generally in tropical wet (evergreen) forest and occasionally in premontane wet forest. The species has a distribution pattern similar to that of *Calathea indecora*. So few collections have been made in the Atlantic coastal forests of Nicaragua and Honduras that its apparent absence from this region may be an artifact of collection. *Calathea warscewiczii* occurs within the forest near stream banks, tree falls, or the forest edge, but rarely grows in open areas. This species is prevalent in the habitat created by cacao plantations, which provides moderate but not dense shade; in these areas the plants are more frequent than in adjacent forests.

Plants of *Calathea warscewiczii* cultivated in Brasil were sent to Bull in England and were described by Hooker (1899) as *C. picta* and cited as "native of Brazil."

Variation: *Calathea warscewiczii* is strikingly uniform throughout its range in qualitative features of habit, inflorescence morphology, and leaf shape and pubescence. The width of the fishtail pattern on the leaf is variable within a population, in some plants it is almost as wide as the leaf itself, but in others it is quite narrow. The markings generally cover 1/2–2/3 of the blade. The plant of *C. warscewiczii* depicted in *Flore des Serres* (Paris) and specimens from cultivation dated within a few years of the original introduction all show a very narrow pattern along the midrib. The lower leaf surface, the leaf sheath, and the stem are almost always dark red purple. However, one individual lacked the purple coloration, and the stem, leaf sheath, and lower leaf surface were light green. The purple color develops only in older plants.

Relationship: The species most closely related to *Calathea warscewiczii* is *C. leucostachys*, which shares a similar inflorescence and flower morphology, though differs from *C. warscewiczii* in the pubescence on the bracts. Both species have short internodes between the basal leaves, giving the plants a rosette appearance. The internode below the subtending leaf (or leaves) is quite long in *C. warscewiczii* and gives it its characteristic habit. This internode is somewhat elongated in *C. leucostachys* also. Both species are rather distantly related to the rest of the closed-flowered species.

Notes: When *Calathea warscewiczii* has almost finished flowering, a new shoot arises in the axil of the subtending leaf. This species commonly occurs in small clumps. These clumps were probably formed by vegetative means and constitute a clone. Of numerous small plants that I uprooted in the field, all were attached to a larger rhizome. The seeds germinate during the latter part of the rainy season. The incidence of seed set in populations in Bocas del Toro was very low and the majority of inflorescences had no capsules at all, and only a few had one or more. The seeds germinate during the latter part of the rainy season.

Near the La Lola Cacao Institute (Milla 28) Limón Province, Costa Rica, a natural hybrid between *C. warscewiczii* and *C. marantifolia* was found by Dr. R. L. Dressler (*H. Kennedy and R. L. Dressler 1390*, US). Three artificial self-pollinations of the hybrid produced swollen ovaries which aborted within seven days.

SUMMARY

A natural group of fifteen primarily Central American species is recognized within the *Scapifoliae*, characterized by the "closed" but not cleistogamous flower and a similar seed and capsule morphology. Fourteen of the fifteen species occur either in Costa Rica or Panamá or both. A single species is restricted to southern México, three are endemic to the Osa Peninsula of Costa Rica, and one species occurs in Panamá and northern South America. Four species are recognized as new, *Calathea lasiophylla, C. longiflora, C. silvicola,* and *C. trichoneura.* A variety, *C. macrosepala* var. *reflexa,* and a subspecies, *C. silvicola* ssp. *tomentosa,* were described.

Flowers of this group are pollinated by long-tongued Euglossine bees in the genus *Eulaema* (four species) and the genus *Euglossa* (three species). Flowers do not open spontaneously but are forced open by the pollinators, which enter and trip the spring mechanism. The flowers are visited solely for nectar. The pollen is transferred in the bud to the back of the style immediately behind the stigma. At anthesis the style is under tension. When it is released by the pollinator, the style snaps forward, the stigma enters the proboscidial fossa and thus contacts the pollen deposited on the insect by the previously visited flower, and the style deposits its pollen in the same place. The stylar movement throws the bee forward against the callose staminode and pins it there momentarily. The bee frees itself and proceeds to feed.

The pollinator/plant specificity is rather one-sided, and the plants are monophilic, pollinated by a single species, or a few closely related species, while the bees are polytrophic, feeding on various, often unrelated, plant species. For each population, usually only a single pollinator species was observed at a given time. The presence of two pollinators on the same population was almost always correlated with high plant densities. The main factor determining the pollinator or pollinators of a plant species is the degree of overlap of the ecological tolerances (ranges) of the pollinators with the plants. Calatheas that occur in a wide variety of habitats have the greatest number of observed pollinators and the least apparent "specificity." Those species with a single observed pollinator species are the forest-dwelling, wet forest species.

There is a striking correlation between the corolla tube length and the habitat type in which the species occur. Species having the shortest tube lengths occur in the driest (semideciduous forest) habitats, usually in the open areas, while those having the longest corolla tubes are in deep, wet forest habitats. Similarly the dry-adapted bees of more open areas have shorter tongue lengths than the forest-dwelling, wet forest species. In addition to the trend from short to long corolla tubes (from open/drier habitats to forest/wet habitats), there is a general decrease in the number of bracts per inflorescence, a decrease in the number of flowers produced per inflorescence per day, an increase in seed size, and a shift from relatively large, high-density populations to populations of fewer, more scattered plants.

The Euglossines also figure importantly as specialized nectar thieves of the closed-flowered Calatheas. Whereas the smaller-sized Eulaemas are the major pollinators of the closed-flowered Calatheas, the larger-sized Eulaemas are commonly robbers. Two modes of robbing were observed: the *E. meriana* type where the flowers are opened but not tripped (and can later be pollinated) and the *E. seabrai* type where the bee trips the flower but avoids the spring style. In the latter type, the flower cannot be pollinated. The type of robbing behavior appears to be species specific; the behavior for each bee species was consistent wherever it was observed. The corolla tube length and the availability of nectar largely determine which plant species are robbed. Because of its manner of robbing, *E. meriana* visits shorter-tubed flowers than do pollinators with similar tongue lengths, but *E. seabrai* visits the same range of tube lengths as do pollinators with similar tongue lengths. The frequency of robbing in a population seems to depend on the size of the plant population in relation to the resident pollinator population. Robbing is most prevalent in high-density plant populations where a sizable portion of flowers remain untouched by the pollinators. The larger robber species require more closely spaced flowers than do the smaller pollinators to make foraging profitable.

The closed flowers represent a specialized, more advanced, floral condition that has evolved from the open-flowered condition. The presence of the outer staminode in the closed-flowered species clearly demonstrates the direction of evolution. The outer staminode, which serves an important attractive function in the open-flowered species, is considerably reduced (occasionally absent) and quite variable in size in the closed-flowered species. Such a reduction or loss of floral structures is characteristic of cleistogamous flowers which the flowers of these species resemble; however, in the closed-flowered species there is no concomitant loss of nectar production—they do not self and therefore must attract pollinators.

The closed flower probably arose as a result of the tendency of Euglossines occasionally to open the buds of open-flowered species prior to their natural opening time. If the amount of nectar produced by a species was slightly greater than that of other species pollinated by Euglossines in the area, the bees would preferentially visit the former earlier in the morning. Thus, a loss of the spontaneous opening of the flower might not have adversely affected the chances of being pollinated. Further, the striking differences in morphology between an unopened and an opened, tripped "closed" flower could have led to greater foraging efficiency for the pollinator, since an untouched flower (with nectar) could be readily distinguished from a visited (nectarless) one at a greater distance than could the tripped and untripped open flowers. Also, robbed flowers are apparent from their morphology, thus the bees learn to associate the closed-flower morphology with a full nectar complement, and their visitation to these flowers might be more constant.

INDEX TO EXSICCATAE

Abbreviation		Species
ds	12.	*C. donnell-smithii*
fo	10.	*C. foliosa*
in	8.	*C. indecora*
ls	7.	*C. lasiophylla*
lt	3.	*C. latifolia*
lu	14.	*C. leucostachys*
lo	13.	*C. longiflora*
mcm	2a.	*C. macrosepala* var. *macrosepala*
mcr	2b.	*C. macrosepala* var. *reflexa*
mr	4.	*C. marantifolia*
ss	9a.	*C. silvicola* ssp. *silvicola*
st	9b.	*C. silvicola* ssp. *tomentosa*
so	1.	*C. soconuscum*
tr	5.	*C. trichoneura*
ve	11.	*C. venusta*
vi	6.	*C. vinosa*
wa	15.	*C. warscewiczii*

Allen, P. H., 861 *lt*; 2219, 2331, 2459, 3964 *ds*; 5340 *vi*; 6277, 6601 *mr*. André, E. F., 462, K590 *lt*. Araque M., J. & Barkley, F. A., 18C748 *mr*. Aristequieta, L., 4307 *lt*. Aristequieta, L. *et al.*, 6280 *lt*. Aviles, S., 15 *mr*.

Bailey, L. H. & Bailey, E. Z., 580 *lt*. Bartlett, H. H. & Lasser, T., 16611, 16619, 16811 *lt*; 16854 *mr*. Blair, N. W., 1016 *fo*. Blum, K. E. & Dwyer, J. D., 2686 *lt*. Brenes, A. M., 211 *lu*; 223 *ds*; 316A *lu*; 326 *ds*. Breteler, F. J., 4174 *lt*. Broadway, W. E., 2359, 7399, 7737 *lt*. Bunting, G. S. & Licht, L., 757 *lu*. Burger, W. C. & Burger, M., 7587 *tr*; 7733 *mcm*; 8427 *tr*; 8440 *wa*; 8465 *mr*; 8495 *fo*; 8498 *in*. Burger, W. C. & Liesner, R., 6244 *tr*; 6263 *fo*; 6828 *tr*; 6929, 6954 *mr*; 7109 *ds*; 7192, 7726 *lo*. Burger, W. C. & Matta, G., 4407 *mr*.

Calderón, S., s.n. *mcm*. Callaway, R., 171 *mcm*. Camp, W. H., E-3783 *mr*. Castañeda, R. R., 3164, 4486 *lt*. Chavelas P., J. *et al.*, s.n. *mcm*. Chickering, A. M., 24 *lt*; 225 *mr*. Cook, O. F. & Doyle, C. B., 582 *st*. Correa, M. D. & Haines, B. L. 247 *mr*. Croat, T. B., 6351 *lt*; 11410 *mr*; 11989 *lt*; 12081 *ds*; 12083, 12222 *mr*; 13199 *ss*; 16886 *lt*. Crueger, H., s.n. *lt*. Cuatrecasas, J., 7375 *mr*; 14243 *lt*; 15162 *mr*; 15208 *ss*; 15673 *mr*; 17233 *lt*. Cuatrecasas, J. & Castañeda, R. R., 24996 *lt*. Cuatrecasas, J. & Willard, L., 26109 *mr*.

Dodge, C. W. & Allen, P. H., 17060 *mr*. Dodge, C. W. & Goerger, V. F., 9413 *in*; 9469, 9497, 9498, 9954, 10019, 10145(GH) *mr*; 10145(CR, F) *ds*. Dodge, C. W. & Thomas, W. S., 4372 *wa*. Dodson, C. H. & McMahon, M., 4339, 5062 *mr*. Dodson, C. H. & Tan, K., 5356 *mr*. Dressler, R. L., 4191 *st*. Dryander, E., 2584 *lt*. Duke, J. A., 3829 *lt*; 5430 *mr*; 12479 *mcm*. Dwyer, J. D., 1236 *lt*; 7170 *mr*. Dwyer, J. D. & Kirkbride, J. H., 7785 *lt*. Dwyer, J. D., *et al.*, 7560 *mcm*.

Ebinger, J. E., 475, 838 *lt*; 1117 *ds*. Eggers, H. F. A. von, 13548 *lt*. Enders, A. R., s.n. *lu*.

Fendler, A., 797 *lt*. Frankie, G. W., 393C *mr*.

García-Barriga, H., 11278 *mr*. Gentry, A. H., 2612 *lt*; 5615 *mcm*; 5831 *mr*. Goll, C. P., 93 *mcm*. Goudot, J., 134 *lt*. Gutiérrez V., G. & Barkley, F. A., 17C215 *lt*.

Hatch, W. R. & Wilson, C. L., 378 *mcm*. Haught, O. L., 2312, 4338 *lt*. Hayes, S., 161 infl. *mr*; 227 *lt*; 821 *mr*. Heyde, A. W. & Lux, E., 3872, 3876, 6256 *mcm*. Hinton, G. B., 13896 *so*. Hodge, W. H., 6509, 6795, 7047 *mr*. Hoffmann, C., 762 *ds*. Holm, R. W. & Iltis, H. H., 253 *mcm*; 345 *wa*;

392 *mr*; 870 *in*; 892 *wa*. Hort. Bot. Berol, s.n. *wa*. Hort. Bot. Imp. Petrop., s.n. *wa*. Hort. Bot. Kew, s.n. *wa*. Hort. Bot. Paris., s.n. *wa*. Hort. Bot. Praga, s.n. *wa*. Hort Bot. Vindob., s.n. *wa*. Horton, O. B., 8650 *mcm*. Humboldt, F. W. H. A. von & Bonpland, A. J. A., 203 *lt* (col. no. *fide* Schumann).

Idrobo, J. M., 1845 *ss*. Idrobo, J. M. & Weber, H., 1370 *lt*.

Jiménez M., A., 1025, 2112 *lu*; 2240 *vi*; 2322 *mr*; 2449 *mcr*; 2477 *ds*; 2870 *wa*; 3024 *vi*. Jiménez, O., s.n. *ds*. Johnson, J. R., 992 *mcm*. Johnston, I. M., 208 *lt*.

Kellerman, W. A., 7318, 7512 *mr*. Kennedy, H. A., 41 *lo*; 57 *ds*; 58 *wa*; 60 *mr*; 63 *mcm*; 267, 269 *mcr*; 306 *ss*; 384 *vi*; 385 *lu*; 386 *mcr*; 406 *wa*; 407 *mr*; 440 *lt*; 454 *lt*; 479 *mr*; 490, 497 *ss*; 567 *mcm*; 640 *ds*; 641 *mr*; 644 *vi*; 647 *lo*; 648 *ls*; 650 *mcr*; 651 *vi*; 751 *ss*; 1138 *ls*; 1142A *lo*; 1158 *ve*; 1165A, 1172, 1173, 1178, 1199, 1221 *lt*; 1234 *ve*; 1515 *ss*; 1521 *ds*; 1534 *mr*; 1539 *ds*; 1540 *ss*; 1559 *fo*; 1563 *in*; 1601 *lo*; 1604 *ds*; 1616 *vi*; 1647 *ds*; 1648 *lu*; 1649 *mcm*; 1661 *lu*; 1666 *mr*; 1686 *tr*; 1936, 1998 *lo*. Kennedy, H. A. & Andrews, R. 1392 *wa*. Kennedy, H. A. & Dressler, R. L., 425 *lt*; 1280, 1284 *wa*; 1285 *mr*; 1290 *in*; 1389 *mr*; 1389A *ve*; 1390 *wa* X *mr*; 1503 *ss*; 1510 *lu*; 2969, 3302 *mr*. Kennedy, H. A. & Dreyfus, D., 389 *in*. Kennedy, H. A. & Foster, R. B., 72, 73 *mcr*; 74 *tr*; 451A *ss*; 662, 689 *mr*. Kennedy, H. A. & Gómez P., L. D., 1571 *tr*. Kennedy, H. A. & Luteyn, J. L., 447 *mr*; 450 *ss*. Kennedy, H. A. & Powell, G., 576 *mr*. Kennedy, H. A. & Ramíerez, W., 1144 *tr*. Kennedy, H. A. & Redemske-Young, J., 1828 *mr*. Kennedy, H. A. & Seifert, F., 1600 *ls*. Kennedy, H. A. *et al.*, 78, 290, 294, 1101, 1128 *ss*; 1157 *lu*; 1381 *mr*; 1386 *wa*; 1461 *mcm*; 1462 *so*; 1465, 1473 *mcm*; 1700 *lt*. Killip, E. P., 3430, 12032, 12081 *lt*; 34724, 35545, 39978 *mr*. Killip, E. P. & Smith, A. C., 14293 *lt*. Kirkbride, J. H. & Duke, J. A. 629 *st*. Koch, R. G., 4880 *mcm*. Køie, M. E., 5323 *mr*. Kupper, W., 142 *mr*; 451 *wa*.

Labr. Químico, 1 *mcm*. Langlassé, E., 275 *so*. Lee, V. H., 64 *lt*. Lent, R. W., 1194 *lu*; 1328 *mr*; 1409 *ve*; 2164 *mcm*; 2202 *tr*. León, J., 751 *ds*; 4206 *mr*. Lévy, P., 1176 *mcm*. Lewis, B. B., 279 *mr*. Lewis, W. H. *et al.*, 122, 128 *mr*; 982 *wa*; 983 *mr*; 2098 *wa*; 3041 *ds*; 3141 *lt*. Liebmann, F. M., 14706 *so*. Long, L. E., 66 *mr*. Luteyn, J. L., 3321 *tr*; 3322 *fo*; 3393 *mcm*. Luteyn, J. L. & Foster, R. B., 1365 *mcm*. Luteyn, J. L. & Kennedy, H. A., 1715 *lt*; 1716 *mr*; 1839 *ss*.

Matuda, E., 16074(F, MEXU, MICH) *mcm*; 16074(MO) *so*; 16592, 18012 *so*; 19346 *mcm*. Matos, F., 1053 *lt*. Maxon, W. R. *et al.*, 7449, 7495 *mcm*. Miranda, F., 1807 *mcm*. Molina R., A., 3400 *mr*; 5440, 5687 *mcm*; 15839 *mr*; 21961 *mcm*. Moore, C. E., 10 *in*.

Nakamura, K., 27(MEXU 117035, 89551, s.n.) *so*; 27(MEXU 89550) *mcm*; 28 *mcm*.

Oliver, R. L. *et al.*, 3666 *ss*. Opler, P. A. 420 *wa*. Ørsted, A. S., s.n. *ds*; s.n. *mr*; 1 *ds*.

Padilla, S. A., 588 *mcm*. Paul, Bro., 103 *lt*. Philipson, W. R. *et al.*, 1345 *lt*. Pittier, H., 1192, 2822 *wa*; 4050 *mr*; 4061, 4129 *lt*; 6746 *mr*; 6753, 6754 *lt*; 8832 *mcm*; 9047 *lt*; 9765 *mr*; 11151, 11850 *lt*; 10363, 13442, 16157 *wa*; 16727 *fo*. Porter, D. M., *et al.*, 4971 *ds*.

Raven, P. H., 21598 *ls*; 21720 *lo*; 21754 *mr*; 21935, 21942 *ds*. Reko, B. P., 3305 *so*. Robinson, J. W. L., 193 *mr*; 345 *ss*. Rooden, J. van & Akkermans, A., 113 *mr*. Rowlee, W. W. & Stork, H. E., 976 *lt*.

Saer, J. 608 *lt*. Sanchez M., M. J. & López, J., 59 *mr*. Schipp, W. A., 8-618, 8-919 *mcm*. Schultes, R. E. & Cabrera, I., 18645 *lt*. Schultze, A., 512 *lt*. Seibert, R. J., 647 *mcm*. Seler, E., 3320 *mr*. Simmonds, N. W., 11, 12, 167 *lt*. Skutch, A. F., 2688 *mcm*; 2778 *ds*. Smith, C. E. *et al.*, 3288 *lt*. Smith, F. D., 199 *lt*. Smith, H. H., 2425 *lt*. Smith, S. G. & Idrobo, J. M., 1517 *lt*. Solís, R. F., 418 *mr*. Standley, P. C., 25164, 25224, 25441, 25709, 25968, 26477, 26676, 26839, 27942, 29065 *lt*; 31392 *mr*; 39500 *tr*; 40179 *ds*; 40927, 52737, 54316, 54928 *mr*; 63549 *mr*; 78199 *mcm*. Standley, P. C. & Torres, R., 51367, 51394, 51706 *tr*. Standley, P. C. & Valerio, J., 44833, 45310 *mr*; 46765 *ds*; 48860, 48865 *in*. Stevens, F. L., 1197 *lt*. Steyermark, J. A., 29590, 51237 *mcm*. Steyermark, J. A. & Allen, P. H., 17060 *mr*. Steyermark, J. A. & Fernández, J. E., 99819 *lt*. Steyermark, J. A. & Rabe, M., 96450 *lt*. Stork, H. E., C '119' *wa*. Sucre, D., 81, 244 *lt*.

Tessene, M. F., 1409 *lu*. Tonduz, A., 8102 *wa*; 8310 *mr*; 8368 *ds*; 9437 *st*; 9775 *mcm*; 11382 *wa*; 11384 *ds*; 12540 *tr*; 13299 *wa*; 13300 *mr*; 14514 *wa*; 14547 *mr*. Tyson, E. L., 1359, 1416 *lt*; 1436 *mr*. Tyson, E. L. & Clewell, A. F., 5916, 5917 *lt*. Tyson, E. L., Godfrey, R. K., *et al.*, 2389 *ds*.

Umaña, E., 44 *wa*. United Fruit Comp., s.n. *wa*; 246 *wa*; 320 *mr*.

Wagner, M., s.n. fasc. 1 *mr*. Wedel, H. von, 209, 476 *in*; 680 *wa*; 712 *in*; 772 *mr*; 903, 1077 *wa*; 1229 *in*; 1270 *wa*; 1328 *mr*; 1586 *wa*; 2102 *in*; 2655 *wa*; 2706 *mr*; 2908 *in*. Wendehake, C., 21 *lt*. Wilbur, R. L. & Stone, D. E., 9700 *lu*. Wilson, P., 100 *mr*. Woods, F. *et al.*, s.n. *lt*. Woodson, R. E. *et al.*, 425, 1669, 1670 *lt*. Woodson, R. E. & Schery, R. W., 1010, 1011 *lt*.

Yuncker, T. G., 4917 *mr*. Yuncker, T. G. *et al.*, 8460 *mr*.

LITERATURE CITED

ALLEN, P. H.

1956. The Rain Forests of Golfo Dulce. University of Florida Press, Gainesville. 417 pp.

BAKER, H. G. and P. D. HURD

1968. Intrafloral ecology. Ann. Rev. Ent. 13: 385-414.

BEARD, J. S.

1944. Climax vegetation in tropical America. Ecology 25: 127-158.

BENNETT, R. D.

1965. Notes on a nest of *Eulaema terminata* Smith (Hymenoptera, Apoidea) with a suggestion of the occurrence of a primitive social system. Insectes Sociaux 12: 81-92.

BENTHAM, G. and J. D. HOOKER

1883. Genera Plantarum 3: 653-654.

BRIAN, A.D.

1957. Differences in the flowers visited by four species of bumblebees and their causes. J. Anim. Ecol. 26: 71-98.

COSTERUS, J. C.

1918. Die uebereinstimmung und der Unterschied in dem Bau der Blumen von *Canna* und derjenigen der Marantaceen. Ann. Jard. Bot. Buitenzorg. II. 15: 59-90.

DELPINO, F.

1869. Breve cenno sulle relazioni biologiche e genealogiche delle Marantaceae. Nuovo Giorn. Bot. Ital. 1: 293-306.

DODSON, C. H.

1966. Ethology of some bees of the tribe Euglossini (Hymenoptera: Apidae). J. Kansas Entomol. Soc. 39: 607-629.

DODSON, C. H. and G. P. FRYMIRE

1961. Natural pollination of orchids. Missouri Bot. Gard. Bull. 49: 133-152.

DRESSLER, R. L.

1967. Why do Euglossine bees visit orchid flowers? Atlas Simpós. Biota Amazônica 5 (Zoologia): 171-180.

1968a. Observations on orchids and Euglossine bees in Panama and Costa Rica. Rev. Biol. Trop. 15: 143-183.

1968b. Pollination by Euglossine bees. Evolution 22: 202-210.

DUCKE, A.

1901. Beobachtungen über Blütenbesuch, Erscheinungszeit etc. der bei Pará vorkommenden Bienen. Zeitschr. Syst. Hym. Dipt. 1: 25-32; 49-51; 60-63.

EICHLER, A. W.

1875. Blüthendiagramme construirt und erläutert. Vol. 1. Leipzig: W. Engelmann.

1883 (1884). Beiträge zur Morphologie und Systematik der Marantaceen. Berlin: Abh. Königl. Akad. Wiss.

ERDTMAN, G.

1952. Pollen Morphology and Plant Taxonomy. Waltham, Mass.: Cronica Botanica Company.

FAEGRI, K. and L. VAN DER PIJL

1966. The Principles of Pollination Ecology. London: Pergamon Press.

GRANT, V.

1949. Pollination systems as isolating mechanisms in angiosperms. Evolution 3: 82-97.

GRIS, A.

1859. Observations sur la fleur des Marantées. Ann. Sci. Nat. IV. 12: 193-219.

HARPER, J. L., P. H. LOVELL, and K. G. MOORE

1970. The shapes and sizes of seeds. Ann. Rev. Ecol. System. 1: 327-351.

HEINRICH, B.

1976. Bumblebee foraging and the economics of sociality. Amer. Scientist 64: 384-395.

HEINRICH, B. and P. H. RAVEN

1972. Energetics and pollination ecology. Science 176: 597-602.

HILDEBRAND, F.
1870a. Botanische Notizen aus einem Briefe von Fritz Müller. (Itajahy, den 7. Dezember 1869.) Bot. Zeitung 28: 273-275.
1870b. F. Delpino's weitere Beobachtungen über die Dichogamie im Planzenreich mit Zusätzen und Illustrationen. Bot. Zeitung 28: 617-625.
HOLDRIDGE, L. R.
1967. Life Zone Ecology, rev. ed. San José, Costa Rica: Tropical Science Center.
HOLTTUM, R. E.
1951. The Marantaceae of Malaya. Gard. Bull. Straits Settlem. (Singapore) 13: 254-296.
HOOKER, J. D.
1865. *Calathea tubispatha.* Bot. Mag. 91: *t. 5542.*
1875. *Calathea leucostachys.* Bot. Mag. 101: *t. 6205.*
1899. *Calathea picta.* Bot. Mag. 125: *t. 7674.*
HORANINOW, P. F.
1862. Prodromus monographiae Scitaminearum. St. Petersburg.
HUMPHREY, J. E.
1896. Development of the seed in the Scitamineae. Ann. Bot. 10: 1-40.
JANZEN, D. H.
1971. Euglossine bees as long-distance pollinators of tropical plants. Science 171: 203-205.
JONKER-VERHOEF, A. M. E. and P. F. JONKER
1957. Marantaceae. *In* A. A. Pulle ed., Flora of Suriname 1(2): 149-208.
KENNEDY, H.
1973. Notes on Central American Marantaceae I. New species and records from Panamá and Costa Rica. Ann. Missouri Bot. Gard. 60: 413-426.
KLOTZSCH, J. F.
1848. Ordo Cannaceae. *In* Richard Schomburgk, Reisen in Britisch-Guiana (in den Jahren 1840-1844). 3: 917-918. Leipzig.
KNUTH, P.
1899. Handbuch der Blütenbiologie. II. Part 2. Leipzig: W. Engelmann.
KOCH, C.
1857. *Maranta, Thalia, Phrynium* und *Calathea.* Bemerkungen im Allgemeinen, ihre Kultur und Beschreibung einiger neuen Arten. Allgem. Gartenz. 21: 141-143.
KOERNICKE, F.
1858. Beiträge zur Kenntniss der in unsern Gärten cultivirten Maranteen. Gartenflora 7: 66-89.
1862. Monographiae Marantearum prodromus. Bull. Soc. Imp. Naturalistes Moscou 35: 1-147.
LAWRENCE, G. H.
1951. Taxonomy of Vascular Plants. New York: Macmillan Co.
LINDLEY, J.
1819. *Maranta zebrina.* Bot. Reg. 5: *t. 385.*
1826a. *Calathea violacea.* Bot. Reg. 12: *t. 962* text sub *t. 961.*
1826b. *Calathea longibracteata.* Bot. Reg. 12: *t. 1020.*
MATUDA, E.
1950. Marantaceas de Chiapas. Estudio de las plantas de Chiapas VIII. Anales Inst. Biol. Univ. Nac. México 21: 319-343.
MEYER, G. F. W.
1818. Primitiae Flora Essequeboensis. Vol. 1, Goettingen: Dietrich.
MOURE, J. S.
1967. A check-list of the known Euglossine bees (Hymenoptera, Apidae). Atlas Simpós. Biota Amazônica 5 (Zoologia): 395-415.
MÜLLER, H.
1883. The Fertilization of Flowers. London: Macmillan & Co.
NEES V. ESENBECK, C. G.
1831. Ueber die Gattungen *Maranta* und *Thalia.* Linnaea 6: 303-342.

OVIEDO Y VALDÉS, G. F. DE.
1851-1855. Historia general y natural de las Indias, islas y tierra-firme del mar océano. José Amador de los Rios, ed. 4 vol. Madrid.
PAI, R. M. and V. D. TILAK
1965. Septal nectaries in the Scitamineae. J. Biol. Sci. 8: 1-3.
PETERSEN, O. G.
1889. Marantaceae. *In* H. G. A. Engler and K. A. E. Prantl, Die Natürlichen Pflanzenfamilien II. 6: 33-43.
1890. Marantaceae. *In* Martius, Flora Brasiliensis III. 3: 81-172.
PERCIVAL, M.S.
1965. Floral Biology. London: Pergamon Press.
PIJL, VAN DER, L.
1960. Ecological aspects of flower evolution. I. Phyletic evolution. Evolution 14: 403-416.
PLUMIER, P. C.
1703. Nova plantarum Americanarum Genera. Paris: J. Boudot.
ROSCOE, W.
1828. Monandrian Plants of the Order Scitamineae. Liverpool.
SCHUMANN, K.
1902. Marantaceae. *In* A. Engler, Das Pflanzenreich. IV. 48: 1-184.
SPRAGUE, E. F.
1962. Pollination and evolution in *Pedicularis* (Scrophulariaceae). Aliso 5: 181-209.
STANDLEY, P. C.
1927. New plants from Central America. J. Wash. Acad. Sci. 17: 250.
1937. Marantaceae. *In* Flora of Costa Rica. Publ. Field Mus. Nat. Hist., Bot. Ser. 18: 191-196.
STANDLEY, P. C. and S. CALDERÓN
1925. Lista Preliminar de Las Plantas de El Salvador. San Salvador: Dutriz.
STANDLEY, P. C. and J. STEYERMARK
1952. Marantaceae. *In* Flora of Guatemala. Fieldiana Bot. 24: 207-221.
STRAW, R. M.
1956. Adaptive morphology of the *Penstemon* flower. Phytomorphology 6: 112-119.
STURTEVANT, W. C.
1969. History and ethnography of some West Indian starches. *In* Ucko, P. J. and G. W. Dimbleby eds., The Domestication and Exploitation of Plants and Animals, 177-199. Chicago: Aldine Publishing Co.
THOMPSON, J. McLEAN
1933. Studies in advancing sterility. Pt. VI. The theory of Scitaminean flowering. Publ. Hartley Bot. Lab. 11: 5-111.
TILAK, V. D. and R. M. PAI
1966. Studies in the floral morphology of the Marantaceae. I. Vascular anatomy of the flower of *Schumannianthus virgatus* Rolfe, with special reference to the labellum. Canadian J. Bot. 44: 1365-1370.
1968. Studies in the floral morphology of the Marantaceae II. Vascular anatomy of the flower in two species of the genus *Phrynium* Willd. Proc. Indian Acad. Sci. 68: 240-249.
TOMLINSON, P. B.
1961. Morphological and anatomical characteristics of the Marantaceae. J. Linn. Soc. Bot. 58: 55-78.
VOGEL, S.
1963. Das sexuelle Anlockungsprinzip der Catasetinen und Stanhopeen-Blüten und die wahre Funktion ihres sogenannten Futtergewebes. Öst. Bot. Zeitschr. 110: 308-337.
1966. Parfumsammelnde Bienen als Bestauber von Orchidaceen und *Gloxinia*. Öst. Bot. Zeitschr. 113: 302-361.
WEBER, H.
1957. Die Wurzelverdickungen von *Calathea macrosepala* Schum. und von einigen anderen monocotylen Pflanzen. Beitr. Biol. Pflanzen Band 34, Heft 1: 177-193.

WOODSON, R. E., JR., and R. W. SCHERY

1942. Marantaceae. *In* Contributions toward a Flora of Panama. VI. Ann. Missouri Bot. Gard. 29: 331-335.

1945. Marantaceae. *In* Flora of Panama. Ann. Missouri Bot. Gard. 32: 81-105.

ZUCCHI, R., S. R. SAKAGAMI, and J. M. F. DE CAMARGO

1969. Biological observations on a Neotropical parasocial bee *Eulaema nigrita*, with a review on the biology of Euglossinae (Hymenoptera, Apidae). A comparative study. J. Faculty Sci., Hokkaido Univ., VI, Zoology 17: 271-380.

Plate 1. "Things are seldom what they seem . . ." Gilbert and Sullivan, *H. M. S. Pinafore*. *Calathea latifolia* being robbed by nectar thief *Eulaema meriana*, Panana Canal Zone.

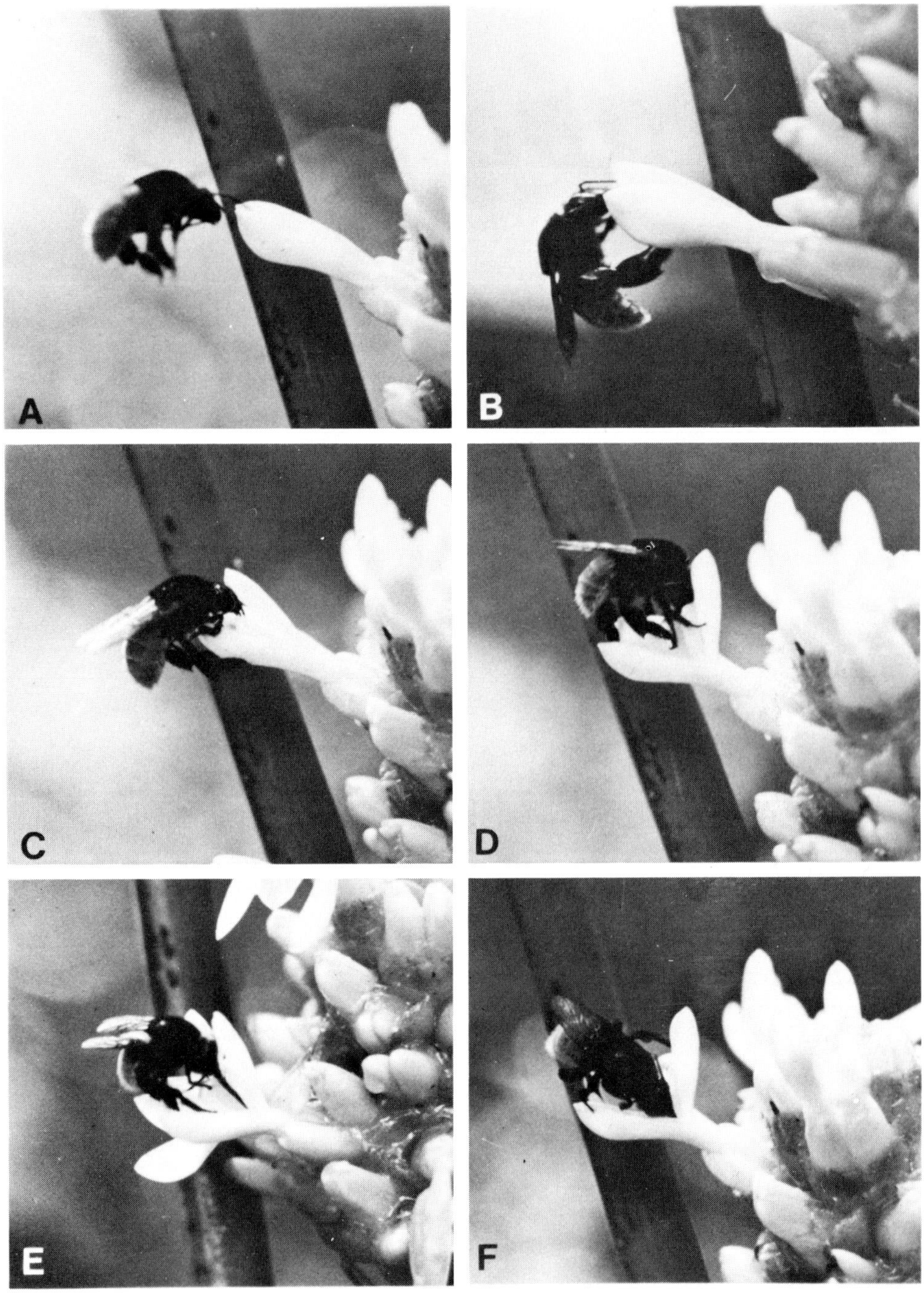

Plate 2. *Eulaema speciosa* pollinating a flower of *Calathea marantifolia* near San Vito de Java, Costa Rica. A. The bee hovers in front of a closed flower. B. The bee chews the tip of the flower and pulls the petals apart. C. The median petal is forced up and away from the callose staminode. D. The bee is thrown forcibly forward by the action of the spring style and is pinned momentarily against the callose staminode. E. The bee reinserts the proboscis in the corolla tube after freeing its head. F. Feeding position.

Plate 3. A. *Eulaema speciosa* feeding on a flower of *Calathea marantifolia.* B. Longitudinal section of an untripped, closed flower. C. Longitudinal section of a tripped flower. Note the expanded basal area of the corolla tube, the nectar holding area, and the constriction in the corolla tube (arrows) just above it.

Plate 4. *Calathea soconuscum* (scale in cm.). A. Habit. B. Inflorescence. C. A tripped flower (right), an unopened flower (center), and at left, top to bottom, the outer staminode, the callose staminode, and the cucullate staminode.

Plate 5. *Calathea macrosepala* var. *macrosepala* (scale in cm.). A. Habit, scale 50 cm. Note the two broad bands on the underside of the leaf. B. Inflorescence. C. An unopened flower (right); a tripped flower (center); at left, cucullate staminode (above) and outer staminode (below). D. *Eulaema cingulata* pollinating a flower of *Calathea macrosepala* var. *macrosepala.* E. Capsules and seeds. Note the swollen base of the capsule, at bottom the persistent calyx abscised from the capsule. F. Root-tubers.

Plate 6. *Calathea macrosepala* var. *reflexa* (scale in cm.). A. Habit, scale 20 cm. high. Note (purple) bands on underside of leaf. B. Inflorescence. Note characteristically reflexed bract margins. C. Flowers. At left an unopened flower; center (top to bottom), outer staminode, callose staminode, and cucullate staminode; at right a tripped flower. D. Capsule with persistent calyx, above; below, capsule and seeds at time of dehiscence. Note swollen basal portion of capsule.

Plate 7. *Calathea latifolia* (scale in cm.). A. Habit. Note bands on underside of upturned leaf. B. Inflorescence, flowers not tripped. C. Underside of leaf showing bands. D. Flowers, tripped and untripped (unopened) with flower parts, top to bottom: outer staminode, callose staminode, petal. E. *Eulaema cingulata* forcing head into flower after opening bud. Uppermost flower has been opened but neither tripped nor pollinated. F. Immature capsule with persistent calyx, mature dehisced capsules (note mealy expanded tissue at base), and seeds on leaf (midrib area).

Plate 8. *Calathea marantifolia* (scale in cm.). A. Habit. B. Inflorescence. Note partially opened, robbed flower on lower left. C. Tripped and unopened flowers on leaf, the callose staminode in the center. D. Variation in the form and size of the outer staminode in a population at Las Cruces, near San Vito de Java, Costa Rica. E. Capsule with persistent calyx, dehisced capsule, and seeds with expanded white aril.

Plate 9. *Calathea trichoneura* (scale in cm.). A. Habit. B. Inflorescence. Note acuminate apex of bracts. C. A tripped and an unopened flower on the leaf surface. D. Close-up of the leaf showing pubescent veins characteristic of this species.

Plate 10. *Calathea vinosa* (scale in cm.). A. Habit. B. Inflorescence. Note unopened flower at lower right, tripped flower above. C. Flowers. Unopened flower at left; center flower showing style and outer staminode; in rightmost flower, note stamen attached to callose staminode; extreme right, callose staminode above. D. Capsule bearing enlarged persistent sepals, and in center, seeds with basal aril.

Plate 11. *Calathea lasiophylla* (scale in cm.). A. Habit. B. Inflorescence. Note unopened flowers. C. Unopened flower at left, tripped flower at right, with outer staminode (above) and callose staminode (below) on undersurface of leaf. D. Seeds and capsule on leaf.

Plate 12. *Calathea indecora* (scale in cm.). A. Habit, scale 20 cm. high. B. Young inflorescence just beginning to flower. C. Old inflorescence. Note flower near base and seedling resulting from a seed germinating *in situ*. D. An unopened flower and a tripped flower on leaf surface. Note eccentric apex of the lateral petal between them. Far right, top to bottom: cucullate staminode, callose staminode, and outer staminode. E. Capsule and seeds.

Plate 13. *Calathea silvicola* ssp. *silvicola* (scale in cm.). A. Habit. B. Inflorescence. Note the tridentate outer staminode of the flower on the right. C. Tripped flowers with callose (right) and outer (left) staminodes on leaf surface. D. Capsules with persistent calyx on leaf surface. Note "pleated" leaf surface.

Plate 14. *Calathea silvicola* ssp. *tomentosa* (scale in cm.). A. Habit. B. Inflorescence with two unopened flowers. C. An unopened and a tripped flower on the leaf surface. Note additional callose structure beside style. D. Close-up view of the abnormal flower. Note the presence of two outer staminodes and the small callose structure adjacent to the malformed callose staminode.

Plate 15. *Calathea foliosa* (scale in cm.). A. Habit. Note the leaves closely clustered atop elongated stem internode. B. Inflorescence. C. An unopened and a tripped flower on leaf surface; at right, top to bottom: callose staminode, outer staminode, and cucullate staminode. D. Capsule with persistent calyx and seeds on leaf surface.

Plate 16. *Calathea venusta* (scale in cm.). A. Habit, scale 25 cm. high. Note clustered spirally arranged leaves on both flowering and juvenile shoots. B. Inflorescence. Note difference between sepals persistent on capsule and those with the flower. C. Tripped flower and capsule with persistent calyx on leaf. Note tridentate outer staminode. D. Capsule with persistent calyx, dehisced capsule, and seeds.

Plate 17. *Calathea donnell-smithii* (scale in cm.). A. Habit. Note that all leaves are cauline. B. Inflorescence with an unopened flower. C. An unopened and a tripped flower on the leaf surface. Note scalloped appearance of leaf.

Plate 18. *Calathea longiflora* (scale in cm.). A. Habit. B. Inflorescence with an unopened flower. C. A tripped and an unopened flower on the leaf. D. Capsule with persistent calyx and seeds. The apex of the sepals turn dark purple.

Plate 19. *Calathea leucostachys* (scale in cm.). A. Habit. Note pale pattern along midrib. B. Inflorescence. C. A tripped flower and an unopened flower on leaf; center left, callose staminode (above) and cucullate staminode (below); center right, characteristically reduced outer staminodes. D. Roots forming in lower portion of stem.

Plate 20. *Calathea warscewiczii* (scale in cm.). A. Habit. Note the rosettelike lower leaves and the two smaller "subtending" leaves of the inflorescence. The fishtail pattern on the leaf is characteristic of this species. B. Inflorescence. Note bract margins color change to purple with age (dark area in photograph). C. A tripped and an unopened flower on the leaf; in the center are the callose staminode, cucullate staminode, outer staminode, and lateral petal. D. Seeds and capsule, at left the abcissed calyx.